# Geospatial Analysis with SQL

A hands-on guide to performing geospatial analysis by unlocking the syntax of spatial SQL

**Bonny P McClain**

‹packt›

BIRMINGHAM—MUMBAI

# Geospatial Analysis with SQL

**Group Product Manager**: Reshma Raman

**Publishing Product Manager**: Heramb Bhavsar

**Content Development Editor**: Priyanka Soam

**Technical Editor**: Rahul Limbachiya

**Copy Editor**: Safis Editing

**Project Coordinator**: Farheen Fathima

**Proofreader**: Safis Editing

**Indexer**: Hemangini Bari

**Production Designer**: Alishon Mendonca

**Marketing Coordinator**: Nivedita Singh

First published: January 2023

Production reference: 4210923

Published by Packt Publishing Ltd.

Grosvenor House

11 St Paul's Square

Birmingham

B3 1RB, UK.

ISBN: 978-1-83508-314-7

www.packtpub.com

*To my geospatial colleagues from around the world: thank you for the warm welcome into this vibrant community—always learning, always curious.*

*To my husband Steve, and boys Harrison and Ryland—your ongoing support means the oblate spheroid to me!*

# Contributors

## About the author

**Bonny P McClain** applies advanced data analytics, including data engineering and geo-enrichment, to discussions of poverty, race, and gender. Racism, class exploitation, sexism, nationalism, and heterosexism all contribute to social inequality. Bonny redefines how we measure these attributes and how we can more accurately identify amenable interventions. Spatial data hosts a variety of physical and cultural features that reveal distribution patterns and help analysts and data professionals understand the underlying drivers of these patterns.

Bonny is a popular conference keynote and workshop leader. Her professional goals include exploring large datasets and curating empathetic answers to larger questions, making a big world seem smaller.

# About the reviewers

**Kate Mai** is a GIS solutions architect who has been in the IT industry for the past 10 years. Her main focuses are web map application product design and development, geospatial data integration and management, and system optimization. She is also interested in what the next big thing may be in the IoT, AI, and GIS.

*Thanks to Lily, my puppy. Without her, the review process would've been done three times faster.*

**Lakshmanan Venkatesan** is a principal consultant at a leading IT services company based out of Houston, Texas. During his career, he has led and contributed to the development of location intelligence software and platforms. He has led digital transformation programs for a large oil company for the past four years. His technical expertise includes advanced web-based mapping application development, scripting, model design, system integration, and DevOps/SRE. Research, teaching, and mentoring are also among his interests.

**Emmanuel Jolaiya** is a software engineer with a background in remote sensing and GIS from the Federal University of Technology, Akure, Nigeria. He has consulted for several leading world organizations, including the World Bank, and currently, he consults for Integration Environment and Energy, a German-based organization, on the **Nigeria Energy Support Programme** (**NESP**), where he uses geospatial technology to support electrification planning. He is a 2020 YouthMappers Research Fellow and Esri Young Scholar. As a young innovator, he is currently building Spatial node, a platform where geospatial professionals can showcase their works and discover opportunities. His current interests include Docker, spatial SQL, DevOps, and mobile and web GIS.

# Table of Contents

# 6

# Building SQL Queries Visually in a Graphical Query Builder    129

# 7

# Exploring PostGIS for Geographic Analysis    153

# 8

# Integrating SQL with QGIS    179

# Preface

This book is for learners interested in using SQL and open source platforms for detecting and quantifying patterns in datasets through data exploration, visualization, data engineering, and the application of analysis and spatial techniques.

At its core, geospatial technology provides an opportunity to explore location intelligence and how it informs the data we collect. First, the reader will see the relevance of **geographic information systems (GIS)** and geospatial analytics, revealing the fundamental foundation and capabilities of spatial SQL.

Information and instruction will be formatted as case studies that highlight open source data and analysis across big ideas that you can distill into workable solutions. These vignettes will correlate with an examination of publicly available datasets selected to demonstrate the power of storytelling with geospatial insights.

Open source GIS, combined with PostgreSQL database access and plugins that expand QGIS functionality, have made QGIS integration an important tool for analyzing spatial information.

## Who is this book for?

This book is for anyone currently working with data (or hoping to someday) who would like to increase their efficiency in analysis and discovery by bringing query and spatial analysis to where their data lives—the database or data warehouse.

Geospatial analysis serves a range of learners across a wide spectrum of skill sets and industries.

This book is mindful of expert data scientists who are just being introduced to geospatial skills, as well as the geospatial expert discovering SQL and analysis for the first time.

## What this book covers

*Chapter 1*, *Introducing the Fundamentals of Geospatial Analytics*, introduces the foundational aspects of spatial databases and data types. This chapter also introduces learners to **structured query language (SQL)**.

*Chapter 2*, *Conceptual Framework for SQL Spatial Data Science – Geometry Versus Geography*, explains how to create a spatial database to enable you to import datasets and begin analysis. You will also learn the fundamentals of writing query-based syntax.

*Chapter 3, Analyzing and Understanding Spatial Algorithms*, shows you how to connect databases created in pgAdmin to QGIS, where you will learn how to join tables and visualize the output by selecting layers and viewing them on the QGIS canvas.

*Chapter 4, An Overview of Spatial Statistics*, covers working with spatial vectors and running SQL queries while introducing you to US Census datasets.

*Chapter 5, Using SQL Functions – Spatial and Non-Spatial*, demonstrates how to use spatial statistics in PostGIS to explore land use characteristics and population data. You will learn how to write a user-defined function and run the query.

*Chapter 6, Building SQL Queries Visually in a Graphical Query Builder*, contains examples of how to access a graphical query builder and begin building more complex frameworks by taking a deeper dive and bringing together skills learned earlier.

*Chapter 7, Exploring PostGIS for Geographic Analysis*, looks at pgAdmin more closely to customize workflows by running SQL queries in pgAdmin and visualizing the output within the geometry viewer.

*Chapter 8, Integrating SQL with QGIS*, moves spatial queries back to QGIS as you work with DB Manager and are introduced to raster data and functions.

## To get the most out of this book

To work with the datasets presented in this book, you will need to download QGIS (I suggest the **long-term-release** (LTR) version, as it is the most stable), pgAdmin, and PostgreSQL.

| Software/hardware covered in the book | Operating system requirements |
|---|---|
| QGIS | Windows, macOS, or Linux |
| pgAdmin | |
| PostgreSQL | |

**If you are using the digital version of this book, we advise you to type the code yourself or access the data resources from the book's GitHub repository (a link is available in the next section). Doing so will help you avoid any potential errors related to the copying and pasting of syntax**

## Download the example code files

You can download the example code files for this book from GitHub at `https://github.com/PacktPublishing/Geospatial-Analysis-with-SQL`. If there's an update to the datasets, it will be updated in the GitHub repository.

Please check the drive link present in the ReadME.md file for the datasets

> **Note**
> The output may differ if the user is downloading the data from the source destination, for similar outputs, user should use drive and GitHub data

# Conventions used

There are several text conventions used throughout this book.

SQL queries: Indicates code words in the text, folder names, filenames, file extensions, pathnames, dummy URLs, user input, and Twitter handles. Here is an example: "To explore the mining areas in Brazil, we can use the ST_Area function."

A block of code is set as follows:

```
SELECT row_number() over () AS _uid_,* FROM (SELECT * FROM
getprotecteda('10160475')
) AS _subq_1_
```

When we wish to draw your attention to a particular part of a query, the relevant lines or items are set in bold: "The synopsis of ST_Within includes the following:"

```
boolean ST_Within(geometry A, geometry B);
```

Any command-line input or output is written as follows:

```
Query returned successfully in 129 msec
```

**Bold**: Indicates a new term, an important word, or words that you see onscreen. For instance, words in menus or dialog boxes appear in **bold**. Here is an example: "In pgAdmin, we can observe these results in **Geometry Viewer**."

> **Tips or important notes**
> Appear like this.

# Get in touch

Feedback from our readers is always welcome.

**General feedback**: If you have questions about any aspect of this book, email us at customercare@ packtpub.com and mention the book title in the subject of your message.

**Errata**: Although we have taken every care to ensure the accuracy of our content, mistakes do happen. If you have found a mistake in this book, we would be grateful if you would report this to us. Please visit www.packtpub.com/support/errata and fill in the form.

**Piracy**: If you come across any illegal copies of our works in any form on the internet, we would be grateful if you would provide us with the location address or website name. Please contact us at copyright@packt.com with a link to the material.

**If you are interested in becoming an author**: If there is a topic that you have expertise in and you are interested in either writing or contributing to a book, please visit authors.packtpub.com.

## Share Your Thoughts

Once you've read *Geospatial Analysis with SQL.*, we'd love to hear your thoughts! Scan the QR code below to go straight to the Amazon review page for this book and share your feedback.

https://packt.link/r/1-835-08314-5

Your review is important to us and the tech community and will help us make sure we're delivering excellent quality content.

# Download a free PDF copy of this book

Thanks for purchasing this book!

Do you like to read on the go but are unable to carry your print books everywhere?

Is your eBook purchase not compatible with the device of your choice?

Don't worry, now with every Packt book you get a DRM-free PDF version of that book at no cost.

Read anywhere, any place, on any device. Search, copy, and paste code from your favorite technical books directly into your application.

The perks don't stop there, you can get exclusive access to discounts, newsletters, and great free content in your inbox daily

Follow these simple steps to get the benefits:

1.  Scan the QR code or visit the link below

https://packt.link/free-ebook/9781835083147

2.  Submit your proof of purchase
3.  That's it! We'll send your free PDF and other benefits to your email directly

# Section 1: Getting Started with Geospatial Analytics

Readers will learn how spatial SQL increases efficiency in working with geospatial databases. Also, prior to analyzing geospatial data, the reader will learn about the frameworks and the distinction between geometry and geography. This is the primary difference between SQL and spatial SQL. They will learn about spatial algorithms and will begin integrating GIS functions into a modern SQL computing environment.

This section has the following chapters:

# 1
# Introducing the Fundamentals of Geospatial Analytics

Understanding where something happened is often the key to understanding why it occurred in the first place. A flood wipes out a village in a remote country 1,000 miles away. If you are thinking geospatially, you are curious about climate forecasting, the demographics of the population, the nature of the soil, the topography of the land, and building footprints or structures. The ability to point to a spot on a map is only a small part of data collection. When you begin to think about roads and distances to impacted villages requiring disaster relief, you would be limited by point-to-point Cartesian geometry. This chapter will help you to understand why.

The process of detecting and quantifying patterns in datasets requires data exploration, visualization, data engineering, and the application of analysis and spatial techniques. At its core, geospatial technology provides an opportunity to explore location intelligence and how it informs the data we collect.

Geospatial information is location data that allows the assessment of geographically linked activities and locations on the Earth's surface. Often called the science of "where," geospatial analysis can provide insight into events occurring within geographic areas such as flooding but also patterns in the spread of wildfires and urban heat islands (increased temperatures in cities compared to rural areas).

Collecting and analyzing geographic information may also include non-spatial data providing opportunities to alter the appearance of a map, based on non-spatial attributes associated with a location. Attributes in **geographic information systems** (**GIS**) refer to data values that describe spatial entities. For example, perhaps an attribute table stores information not only on the building polygon footprint but also indicates that the building is a school or residence.

On your map, you may be interested in viewing wastewater treatment plants or buildings complying with green energy requirements within a city or neighborhood boundary. Although this information is non-spatial, you can view and label information associated with specific locations. You are able to direct map applications to render a map layer based on a wide variety of features.

As you learn how to query your data with **Structured Query Language** (**SQL**), the advantages will include flexibility in accessing and analyzing your data. This chapter will introduce SQL and expand upon concepts throughout the remaining chapters.

Before we explore the syntax of SQL, let's introduce or review a few concepts unique to geospatial data.

In this chapter, we will cover Installation: pgadmin, postgreSQL, QGIS and will cover the following topics:

- Spatial databases
- **Spatial reference identifiers** (**SRIDs**)
- Understanding geospatial data types
- Exploring SQL

First, let's become familiar with a few characteristics of geospatial data. You will set up working environments later in the chapter.

## Technical requirements

The following are the technical requirements for this chapter:

- PostgreSQL installation for your operating system (macOS, Windows, or Linux)
- PostGIS extension
- pgAdmin
- **QGIS**
- GitHub link at `https://github.com/PacktPublishing/Geospatial-Analysis-with-SQL`

## Spatial databases

Geospatial data is vast and often resides in databases or data warehouses. Data stored in a filesystem is only accessible through your local computer and is limited to the speed and efficiency of your work space. Collecting and analyzing geographic information within storage systems of relationally structured data increases efficiency.

Spatial databases are optimized for storing and retrieving your data. Data stored in databases are accessed through client software. The client is a computer or hostname (where the database is located).

Next, the server listens for requests on a port. There are a variety of ports, but with PostgreSQL, the default is 5432—but more on that later. The final pieces of information are about security and accessibility. You will select a username and password to access the database. Now, the client has all it needs to access the database. The data within a database is formatted as tables and resides with the host either locally or in an external server linked to a port that listens for requests. This book will focus on local instances with a mention or two about other options in later chapters.

SQL is a standardization for interacting with databases that abides by **American National Standards Institute (ANSI)** data standards. Additionally, SQL has also been adopted as the international standard by the **International Organization for Standardization (ISO)**, the **International Electrotechnical Commission (IEC)**, and the **Federal Information Processing Standard (FIPS)**.

## SRIDs

Each spatial instance or location has an SRID due to the earth being a non-standard ellipsoid. We talk about the surface of the earth, but what does this mean exactly? Do we plunge to the depths of the ocean floor or the highest mountains? The surface of the earth varies depending on where you are on the surface of the earth.

**Spatial reference systems (SRS)** allow you to compare different maps. If they are superimposable, they have the identical SRS. The **European Petroleum Survey Group (EPSG)** is the most popular.

*Figure 1.1* visually demonstrates the discrepancy in mapping distances on the surface of the earth. Rotational forces at the poles flatten the earth, creating bulging at the equator. This phenomenon complicates the accurate calculation of the distance between two points located at different latitudes and longitudes on the Earth's surface. The coordinate system helps to standardize spatial instances:

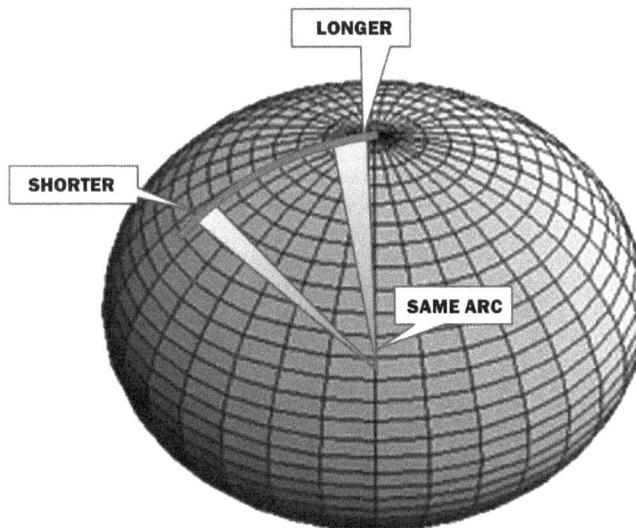

Oblate Ellipsoid (Correct)
Source: Jan Van Sickle

Figure 1.1 – Ellipsoid shape of the earth

When working with location data, the use of reference ellipsoids helps to specify point coordinates such as latitude and longitude, for example. The **Global Positioning System** (**GPS**) is based on the **World Geodetic System** (*WGS 84*). You will notice different IDs based on the types of datasets you are working with—EPSG *4326* is for *WGS 84*. *WGS 84* is the most used worldwide. A spatial column within a dataset can contain objects with different SRIDs, however, only spatial instances with the same SRID can be used when performing operations with SQL Server spatial data methods. Luckily, PostgreSQL will raise an exception when an incompatibility is detected.

SRIDs are stored in a Postgres table visible in the public schema named `spatial_ref_sys`. You can discover all of the SRIDs available within PostGIS (the spatial extension for PostgreSQL) with the following SQL query:

```
SELECT * FROM spatial_ref_sys
```

If you look down the column in *Figure 1.2*, you will notice `EPSG 4326` but also a sample of other SRIDs available:

| | srid<br>[PK] integer | auth_name<br>character varying (256) | auth_srid<br>integer | srtext<br>character varying (2048) |
|---|---|---|---|---|
| 244 | 4605 | EPSG | 4605 | GEOGCS["St. Kitts 1955",DATUM["St_Kitts_1955",SPHEROID["Clarke 1880 (RGS)",6378249.145,293.465,AUTHORITY["... |
| 245 | 4316 | EPSG | 4316 | GEOGCS["Dealul Piscului 1930",DATUM["Dealul_Piscului_1930",SPHEROID["International 1924",6378388,297,AUTHOR... |
| 246 | 4317 | EPSG | 4317 | GEOGCS["Dealul Piscului 1970",DATUM["Dealul_Piscului_1970",SPHEROID["Krassowsky 1940",6378245,298.3,AUTHO... |
| 247 | 4318 | EPSG | 4318 | GEOGCS["NGN",DATUM["National_Geodetic_Network",SPHEROID["WGS 84",6378137,298.257223563,AUTHORITY["EP... |
| 248 | 4319 | EPSG | 4319 | GEOGCS["KUDAMS",DATUM["Kuwait_Utility",SPHEROID["GRS 1980",6378137,298.257222101,AUTHORITY["EPSG","701... |
| 249 | 4322 | EPSG | 4322 | GEOGCS["WGS 72",DATUM["WGS_1972",SPHEROID["WGS 72",6378135,298.26,AUTHORITY["EPSG","7043"]],TOWGS84[... |
| 250 | 4324 | EPSG | 4324 | GEOGCS["WGS 72BE",DATUM["WGS_1972_Transit_Broadcast_Ephemeris",SPHEROID["WGS 72",6378135,298.26,AUT... |
| 251 | 4326 | EPSG | 4326 | GEOGCS["WGS 84",DATUM["WGS_1984",SPHEROID["WGS 84",6378137,298.257223563,AUTHORITY["EPSG","7030"]],A... |
| 252 | 4463 | EPSG | 4463 | GEOGCS["RGSPM06",DATUM["Reseau_Geodesique_de_Saint_Pierre_et_Miquelon_2006",SPHEROID["GRS 1980",63781... |
| 253 | 4470 | EPSG | 4470 | GEOGCS["RGM04",DATUM["Reseau_Geodesique_de_Mayotte_2004",SPHEROID["GRS 1980",6378137,298.257222101,... |
| 254 | 4475 | EPSG | 4475 | GEOGCS["Cadastre 1997",DATUM["Cadastre_1997",SPHEROID["International 1924",6378388,297,AUTHORITY["EPSG",... |
| 255 | 4483 | EPSG | 4483 | GEOGCS["Mexico ITRF92",DATUM["Mexico_ITRF92",SPHEROID["GRS 1980",6378137,298.257222101,AUTHORITY["EP... |
| 256 | 4490 | EPSG | 4490 | GEOGCS["China Geodetic Coordinate System 2000",DATUM["China_2000",SPHEROID["CGCS2000",6378137,298.2572... |
| 257 | 4555 | EPSG | 4555 | GEOGCS["New Beijing",DATUM["New_Beijing",SPHEROID["Krassowsky 1940",6378245,298.3,AUTHORITY["EPSG","702... |
| 258 | 4558 | EPSG | 4558 | GEOGCS["RRAF 1991",DATUM["Reseau_de_Reference_des_Antilles_Francaises_1991",SPHEROID["GRS 1980",637813... |
| 259 | 4600 | EPSG | 4600 | GEOGCS["Anguilla 1957",DATUM["Anguilla_1957",SPHEROID["Clarke 1880 (RGS)",6378249.145,293.465,AUTHORITY["... |

Figure 1.2 – PostGIS SRID

*Figure 1.3* shows the SRID for data in the `NYC` database. These are two distinct datasets, but you can see how variation in SRID might be a problem if you would like to combine data from one table with data in another. When working with a **graphical user interface** (**GUI**) such as QGIS, for example, SRIDs are often automatically recognized.

Data output    Messages    Notifications

| f_table_catalog character varying (256) | f_table_schema name | f_table_name name | f_geometry_column name | coord_dimension integer | srid integer | type character varying (30) |
|---|---|---|---|---|---|---|
| 1 | nyc | public | nyc_census_... | geom | 2 | 26918 | MULTIPOLYGON |
| 2 | nyc | public | nyc_homicides | geom | 2 | 26918 | POINT |
| 3 | nyc | public | nyc_neighbor... | geom | 2 | 26918 | MULTIPOLYGON |
| 4 | nyc | public | nyc_streets | geom | 2 | 26918 | MULTILINESTRING |
| 5 | nyc | public | nyc_subway_... | geom | 2 | 26918 | POINT |
| 6 | nyc | public | geometries | geom | 2 | 0 | GEOMETRY |
| 7 | nyc | public | manhattan_g... | geom | 2 | 4326 | MULTIPOLYGON |
| 8 | nyc | public | MN_lApp | geom | 2 | 2263 | MULTIPOLYGON |
| 9 | nyc | public | MN_LCpp | geom | 2 | 2263 | MULTIPOLYGON |
| 10 | nyc | public | MN_plApp | geom | 2 | 2263 | MULTIPOLYGON |
| 11 | nyc | public | geo_export_2... | geom | 2 | 0 | MULTIPOLYGON |

Figure 1.3 – SRID for NYC database

You can check the geometry and SRID of the tables in your database by executing the command in the query that follows. Don't worry if this is your first time writing SQL queries—these early statements are simply for context. The dataset included here is a standard practice dataset, but imagine if you wanted to combine this data with another dataset in a real-world project:

```
SELECT * FROM geometry_columns;
```

Next to the `srid` column in *Figure 1.3*, you can see geometry types. The next section will introduce you to the different geometries.

## Understanding geospatial data types

To understand how to work with geospatial data and the functions available to answer data questions, we will explore different data types. Geospatial geometries  are represented by polygons, points, and lines that indicate features and attributes such as census blocks, subway station locations, and roadways. They are vector models as they are representative of real-world areas (polygons), locations (points), or linear data such as transportation networks or drainage networks (lines).

Collectively, we call these different geometries vector data. Think of two-dimensional space as *x*- and *y*-coordinates when thinking about a specific geometry. Geography represents data on a round-earth coordinate system as latitude and longitude. I tend to use geometry as the convention across the board but will clarify when relevant. You will see actual data in this chapter, but it is simply illustrative. In later chapters, you will use a data resource, but until then, if you would like to explore, the data in this chapter can be found at *NYC Open Data*: `https://opendata.cityofnewyork.us/`.

*Figure 1.4* displays the output of a query below written to explore the geometry of our data. In spatial queries, geometry is simply another data type in addition to the types we might be familiar with—integer, string, or float. This becomes important when exploring non-spatial data such as demographics. When performing analyses, the data type is relevant as only integer and floating numbers can be used with mathematical functions. Not to worry—there are options to transform numerical properties into integers when needed. You are also able to identify where your table is located (schema) and the name of your geometry column (you will need this information when referring to the geometry in a table). Additional information visible includes the dimensions (two dimensions), the SRID values, and the data type.

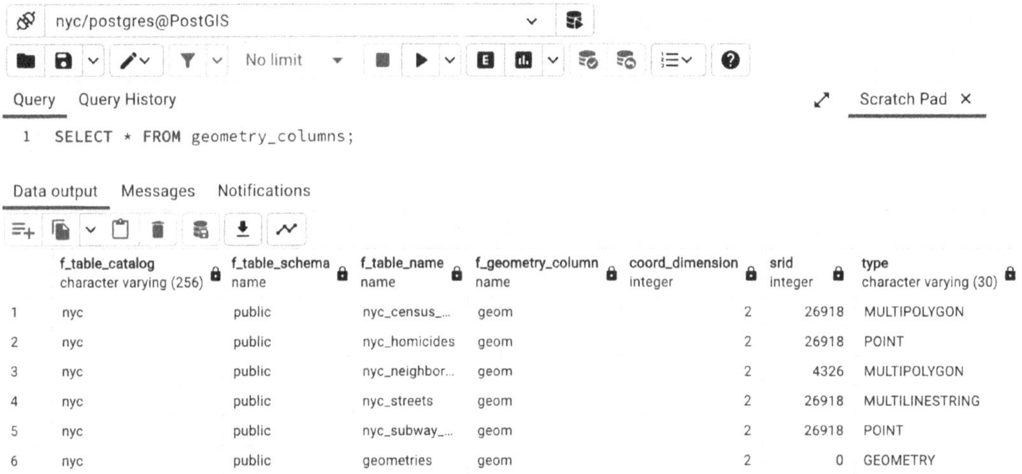

| | f_table_catalog<br>character varying (256) | f_table_schema<br>name | f_table_name<br>name | f_geometry_column<br>name | coord_dimension<br>integer | srid<br>integer | type<br>character varying (30) |
|---|---|---|---|---|---|---|---|
| 1 | nyc | public | nyc_census_... | geom | 2 | 26918 | MULTIPOLYGON |
| 2 | nyc | public | nyc_homicides | geom | 2 | 26918 | POINT |
| 3 | nyc | public | nyc_neighbor... | geom | 2 | 4326 | MULTIPOLYGON |
| 4 | nyc | public | nyc_streets | geom | 2 | 26918 | MULTILINESTRING |
| 5 | nyc | public | nyc_subway_... | geom | 2 | 26918 | POINT |
| 6 | nyc | public | geometries | geom | 2 | 0 | GEOMETRY |

Figure 1.4 – Discovering the geometric types in your data

In the `type` column in *Figure 1.4*, we see three unique types: `MULTIPOLYGON`, `POINT`, and `MULTILINESTRING`. Geospatial data is most often represented by two primary categories: **vector data** and **raster data**.

Roadways are visible in *Figure 1.5* as lines or multiline strings. These are a collection of `Line-String` geometries:

Figure 1.5 – NYC boroughs with shapes and lines

*Figure 1.6* displays the five NYC boroughs with neighborhoods shown as polygons. The points represent subway stations. These are a few examples of how vector geometries are displayed. Most of these features are customizable, as we will see when exploring layer options and symbology in QGIS.

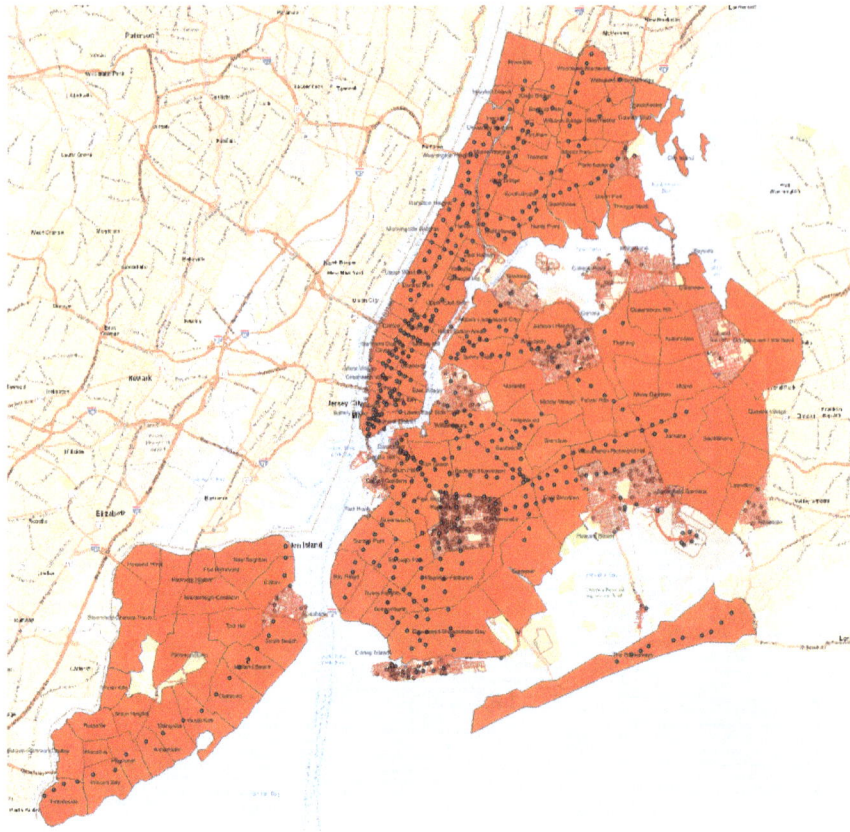

Figure 1.6 – Map of NYC visualizing points, lines, and polygons

Another data type we will be exploring later is raster data. Although PostGIS has raster extensions and functions available, you will need to visualize the output in QGIS. For now, let's understand the characteristics of raster data.

## Raster models

Raster data is organized as a collection or grid. Each cell contains a value. Bands are collected as boolean or numerical values coordinating per pixel. Briefly, multidimensional arrays of pixels or *picture elements* in images are basic units of programmable information. What is measured is actually the intensity of the pixel within a band. The data is in the bands. The intensity varies depending on the relative reflected light energy for that element of the image. Raster data is often considered continuous data. Unlike the points, lines, and polygons of vector data, raster data represents the slope of a surface, elevation, precipitation, or temperature, for example. Analyzing raster data requires advanced analytics that will be introduced in *Chapter 8, Integrating SQL with QGIS*. *Figure 1.7* is displaying Cape Town, South Africa as a **digital elevation model(DEM)**. The darker areas represent lower elevations:

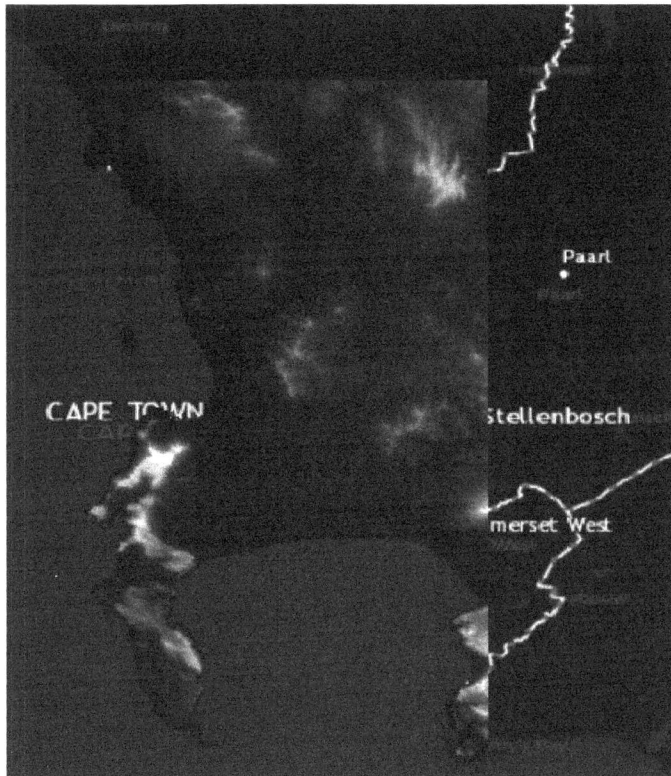

Figure 1.7 – Raster image of Shuttle Radar Topography Mission (SRTM)
elevation and digital topography over Cape Town, South Africa

The complex composition in *Figure 1.7* is representative of raster images and the detailed gradations visible in land use, scanned maps or photographs, and satellite imagery. Raster data is often used as a basemap for feature layers, as you will see in QGIS.

Let's use a few simple SQL queries to understand geographical data. PostgreSQL is the open source **relational database management system** (**RDBMS**) you will explore in addition to PostGIS—the spatial extension—and QGIS, an open source GIS application. You may be familiar with commercial offerings such as Oracle, SQL Server, or SQLite. Open source PostgreSQL has the PostGIS spatial extension, which easily transforms data for different projections and works with a wide variety of geometries.

## Exploring SQL

Let's get a feel for the capabilities and advantages of working with datasets in a spatial database. SQL is described as a declarative programming language. If you have experience in a coding environment, this might be a fairly big departure.

In a declarative programming language, you write what you want to do, and the program figures it out. SQL is a query language that sends queries to manipulate and retrieve data from the database. In an **object-oriented programming (OOP)** language such as Python, for example, you need to specify system calls incorporating libraries and various packages. Although Python's syntax is readable, a familiarity with built-in data structures, powerful libraries, frameworks, and a relative commitment to scripting expertise is required. PostgreSQL includes extensions to write procedures in additional programming languages such as Python and R, for example, but the focus here is on SQL and spatial SQL.

In *Figure 1.8*, you will see a simple SQL statement. Statements have clauses, (SELECT, FROM, or WHERE), expressions, and predicates. Although capitalization is a common practice to assist in readability, SQL is not case-sensitive.

First, let's examine the statement in more detail. The data is from the **Department of Environmental Protection (DEP)** publicly available data (https://data.cityofnewyork.us/City-Government/DEP-s-Citywide-Parcel-Based-Impervious-Area-GIS-St/uex9-rfq8). The objective is to examine NYC-wide parcel-based impervious surfaces. In geospatial analysis, the structure and characteristics of land use, land cover, and surfaces are often explored. Impervious surfaces describe hard surfaces—primarily man-made—that prevent rainwater from reaching groundwater or flowing into waterways and contribute to increased surface temperatures in urban cities. Asphalt parking lots, concrete, and compacted soil in urbanized development are familiar examples.

## SELECT statements

Entering SELECT * FROM public."table name" in the code displays all of the columns in the table. The asterisk saves time as you will not need to ask for all of the columns by name. They will all be returned with the query. Looking at the query in the pgAdmin console in *figure 1.8*, FROM MN_pIAPP after the SELECT statement is referring to the table of percent impervious surfaces in the borough of Manhattan in NYC.

Often when referring to your table in a SQL query, double quotes will be needed with single quotes referring to the values in a column. When you use lowercase as the default, there is an option to omit the double quotes, for example, renaming the table as mn_piapp.

## WHERE statements

The WHERE statement introduces an expression, to return values less than 47.32. The statement will return values for surfaces less than 47.32% impervious. If the predicate is True, the value will be returned in the table. A predicate evaluates which rows are returned in a specific query. *Figure 1.8* illustrates the impervious surfaces in the geometry viewer in pgAdmin:

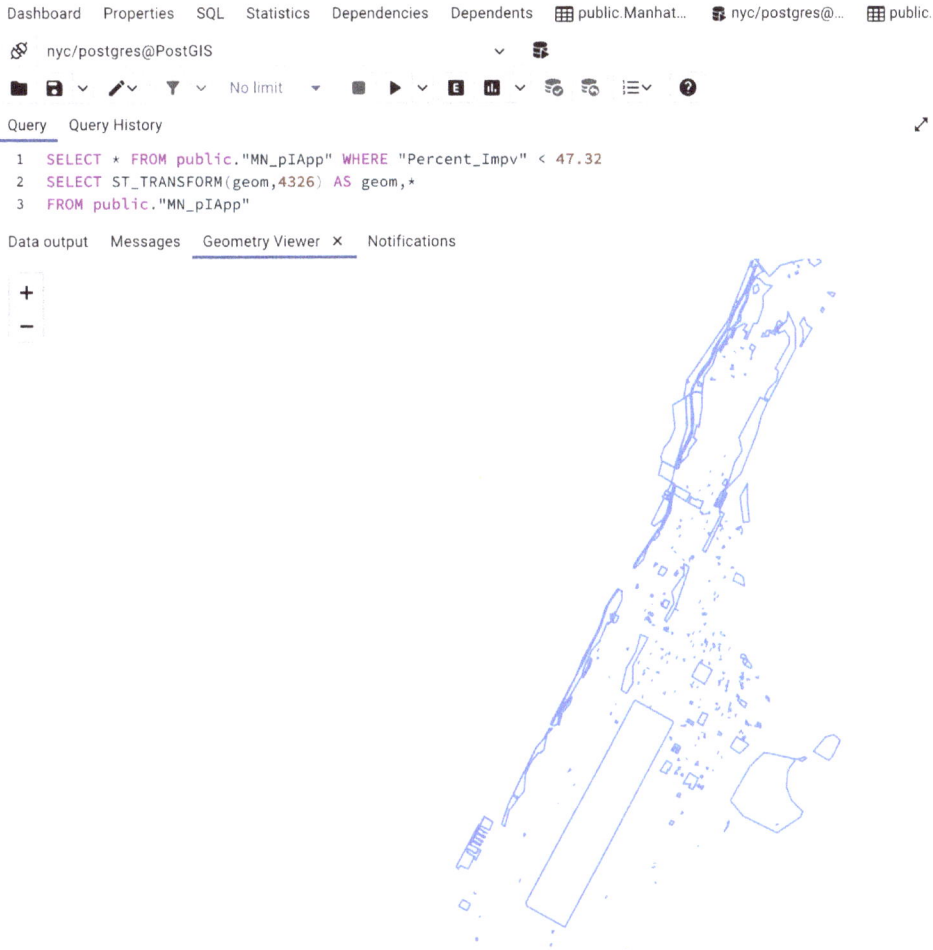

Figure 1.8 – Viewing Manhattan impervious surfaces in Geometry Viewer in pgAdmin

The impervious surfaces are highlighted in the graphic viewer. The output is surfaces less than 47.32% impervious. Change the percentage, and the output will update once you run the following code and view the previous graphical image:

```
SELECT * FROM public."MN_pIApp" WHERE "Percent_Impv" < 47.32
```

The polygons are visible, but what about our basemap? In our GUI, pgAdmin (introduced next in the chapter), we are able to see a basemap if we transform the coordinates to the 4326 SRS or SRID. This is because of a reliance on *OpenStreetMap projected in 3857, and it will also yield a basemap if this is the SRID of your data* (https://www.openstreetmap.org/#map=5/38.007/-95.844). QGIS is a far more powerful visualization tool, and you will explore the PostGIS integration in later chapters. This may be updated to include other projections by the publication of the book so be on the look-out.

`ST_Transform` is a SQL expression allowing you to assign a different SRID once we have the SRID changed to `4326`:

```
SELECT ST_Transform (geom,4326) AS geom,*
  FROM public."MN_pIApp" WHERE "Percent_Impv" < 47.32
```

We are able to view the basemap in **Geometry Viewer** in pgAdmin, shown in *Figure 1.9*:

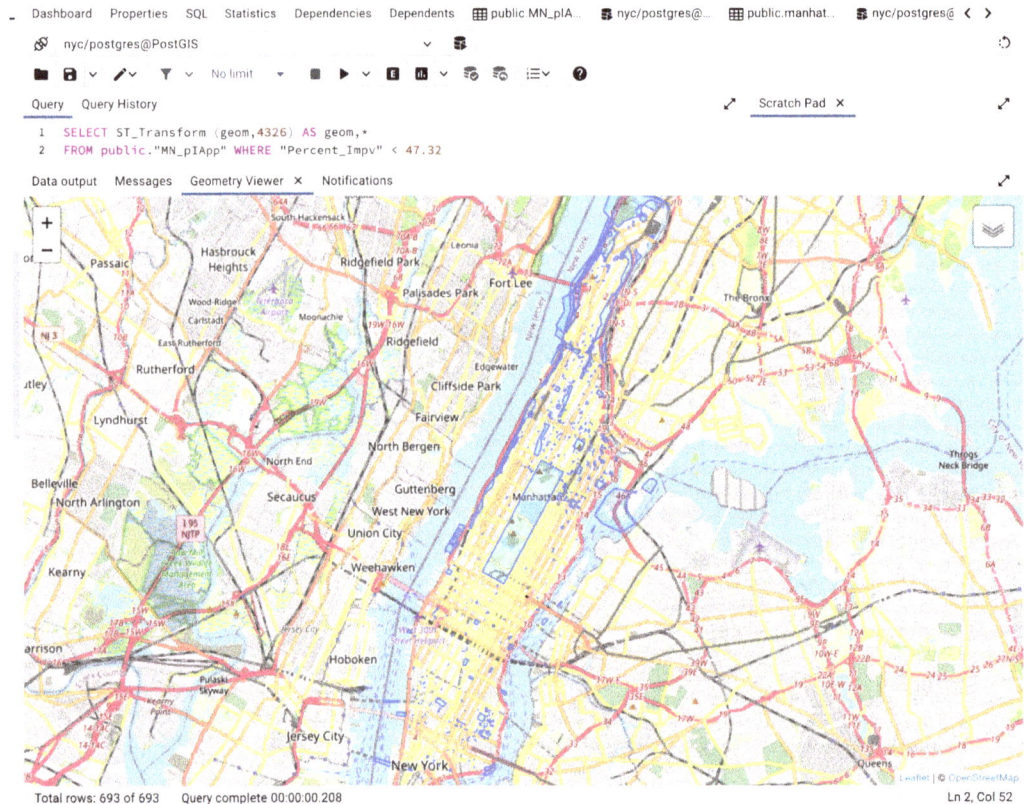

Figure 1.9 – Viewing Manhattan impervious surfaces in Geometry Viewer in pgAdmin with a basemap

The rapidly expanding field of geospatial analytics is evolving across a wide variety of disciplines, specialties, and even programming languages. The potential insights and deeper queries will likely be at the forefront of skill acquisition; however, it is critical that you first explore fundamental geospatial techniques regardless of your preferred computer language or even if you currently aren't working within a particular coding practice.

This book will assume no prior knowledge in either programming or geospatial technology, but at the same time offers professionals anywhere the ability to dive deeper into detecting and quantifying patterns visible through data exploration, visualization, data engineering, and the application of analysis and spatial techniques.

At its core, geospatial technology provides an opportunity to explore location intelligence and how it informs the data we collect. To get started, you will first need to download a few tools.

### Installation of PostgreSQL

PostgreSQL is an open source relational database server with multiple ways to connect. PostgreSQL is also known as Postgres, and I will be referring to the database server as Postgres. Remember—"client" is another word for computer or host. You can access Postgres with a command line in terminal for macOS or in the command line in Windows, through programming languages such as Python or R, or by using a GUI. You will be running Postgres locally and will need both the client and the server (Postgres).

Installation instructions here are specific for macOS but are also pretty straightforward for other operating systems. Download from `https://www.postgresql.org/` to select system-specific options, as shown in *Figure 1.10*:

# Downloads ⬇

## PostgreSQL Downloads

PostgreSQL is available for download as ready-to-use packages or installers for various platforms, as well as a source code archive if you want to build it yourself.

### Packages and Installers

Select your operating system family:

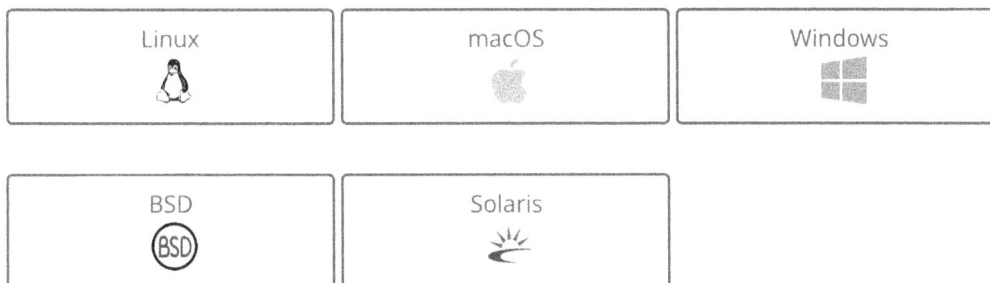

| Linux | macOS | Windows |
| --- | --- | --- |
| BSD | Solaris | |

Figure 1.10 – PostgreSQL download options

Selecting your operating system of choice will yield a series of options. The `Postgres.app` installation for macOS also includes PostGIS, which is the spatial data extension. The examples that follow do not utilize the interactive installer. As I am working on a late model operating system, the potential incompatibilities down the road pointed toward using `Postgres.app`. There are advantages either way, and you can make a choice that works best for your workflow.

When I first began working with PostgreSQL, I noticed many resources for Windows-compatible installations—they are everywhere—but what was lacking, it seemed, was instruction for those of us working with macOS. The focus here will be primarily on macOS because there is not a compatible import function directly for macOS at the time of this printing.

See the following tip box for the **EnterpriseDB (EDB) PostgreSQL installer** information and other downloads selected for the installer option.

> Tip
>
> The installer option will download a setup wizard to walk through the steps of the installation. EDB includes additional extensions to simplify the use of PostgreSQL but for now, download the software onto your local computer.

Windows users and even macOS operating systems can opt for the installation wizard. Follow these steps to set up the installation:

1. Navigate to the option shown in *Figure 1.11* if you are selecting this option:

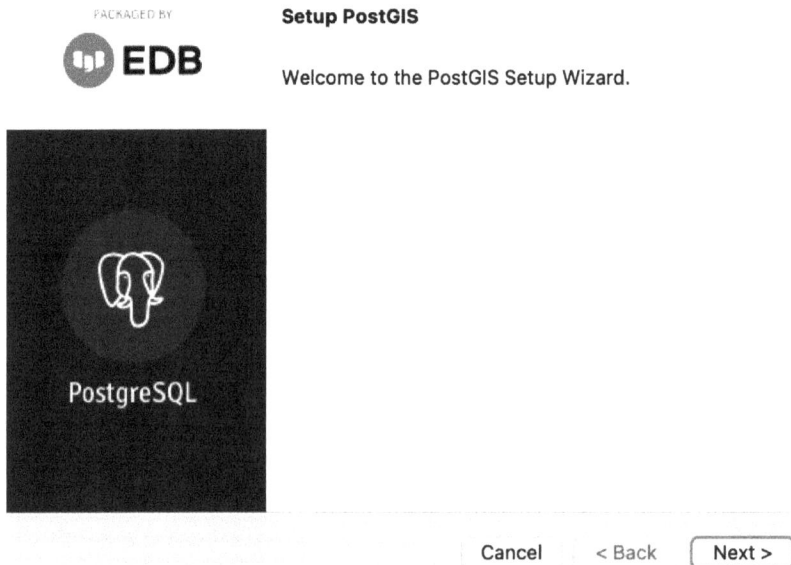

Figure 1.11 – EDB installer for PostgreSQL

2. Stack Builder simplifies downloading datasets and other features we will explore. You can see it being launched in *Figure 1.12*:

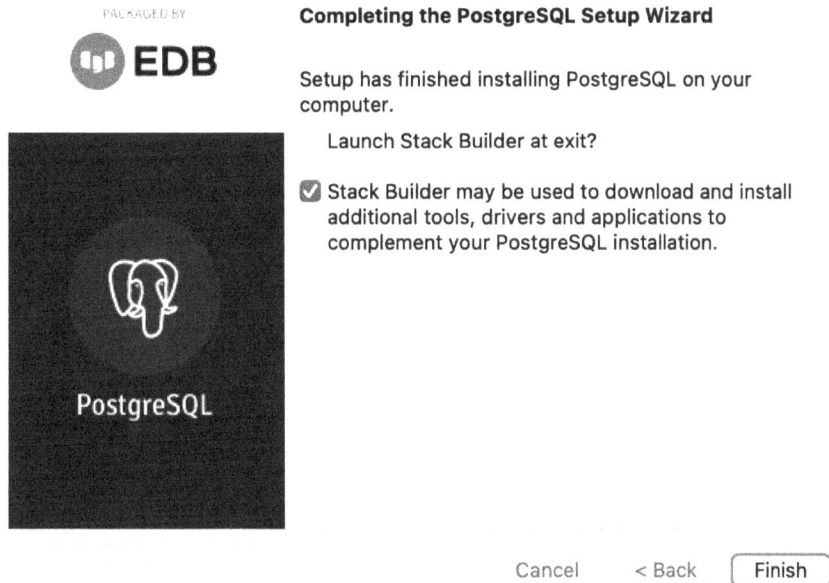

Figure 1.12 – Launching Stack Builder located inside the PostgreSQL application folder

Depending on your setup options or work environment, select the components you would like to download, as given in *Figure 1.13*.

3. Select the four default components, as shown in *Figure 1.13*:

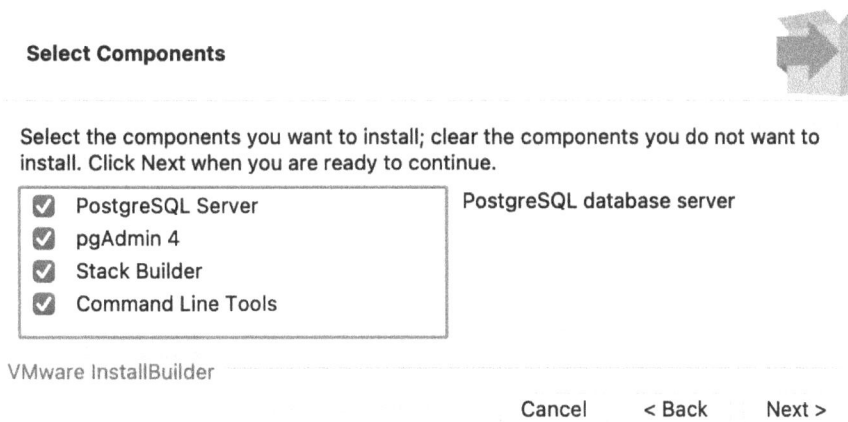

Figure 1.13 – Default components in the installation wizard

Windows has an extensive presence for guiding the installation and a separate installer that includes pgAdmin for managing and developing PostgreSQL, and Stack Builder. Stack Builder is also available in the interactive installer for macOS, but the utility may depend on which version of macOS you have installed.

Unless you are connecting to an existing server or host, localhost should be entered, as seen in *Figure 1.14*. Regardless of your selection of operating system, the default port is 5432. This setting can be changed, but I would suggest keeping the default unless there is a compelling reason to create a different port—for example, if it is already in use. The superuser is postgres. This identity is a superuser because you will have access to all of the databases created. You would customize this if different databases had different authorization and access criteria. Select your password carefully for the database superuser. It is a master password that you will need again.

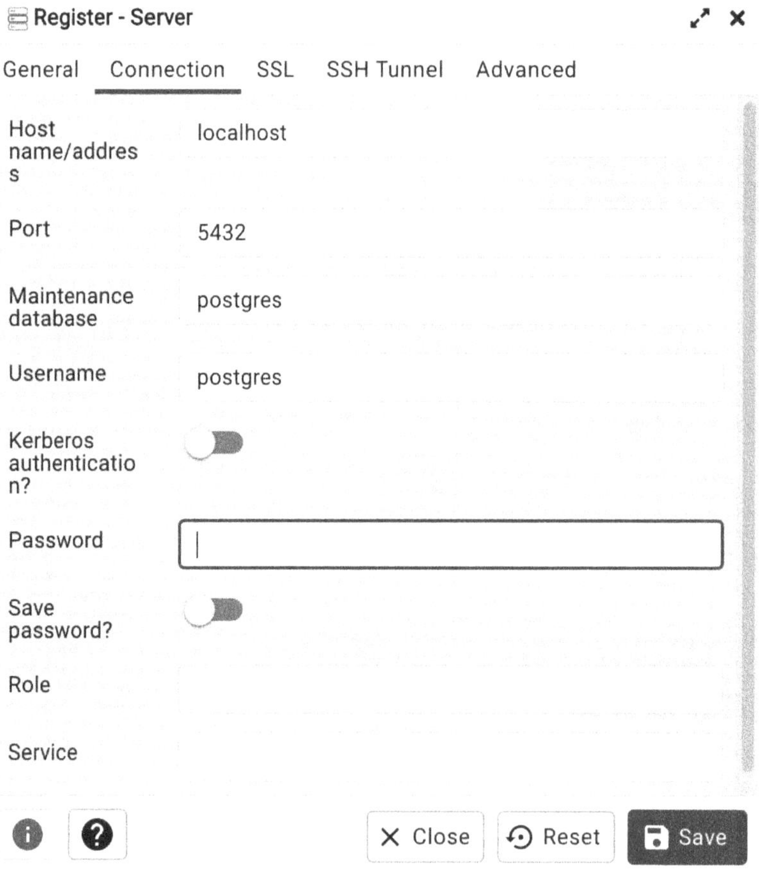

Figure 1.14 – Creating a server instance in Postgres

Now that you have a database server installed locally (downloaded to your computer), you will need a GUI management tool to interact with Postgres. Only a single instance of Postgres is able to run on each port, but you can have more than one open connection to the same database.

## Installation of pgAdmin

pgAdmin is the GUI selected as a graphical interface or **command-line interface** (**CLI**) for writing and executing SQL commands. This is by no means the only option, but pgAdmin works across operating systems and is available for download here: `https://www.pgAdmin.org/`.

Follow the next steps to complete the installation:

1.  Drag the **pgAdmin** icon to the `applications` folder, and voilà! Do not rename the file.

2.  *Figure 1.15* shows the dashboard you will see once you download the software:

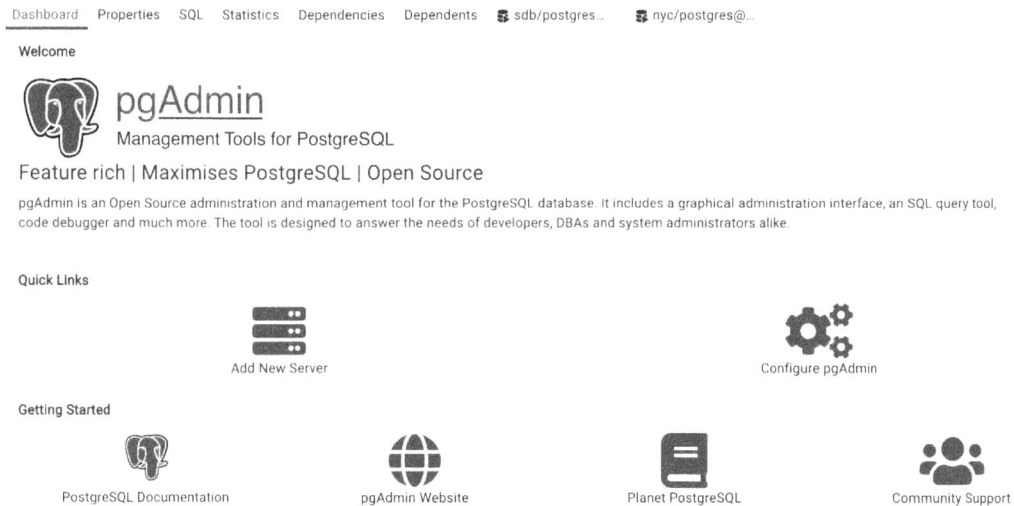

Figure 1.15 – pgAdmin web application for managing PostgreSQL

3.  After logging in to pgAdmin, you will add a server.

4.  Next, you will need a name for your server, as shown in *Figure 1.16*. You have the ability to add servers as needed. The default settings are fine, and the default is Postgres:

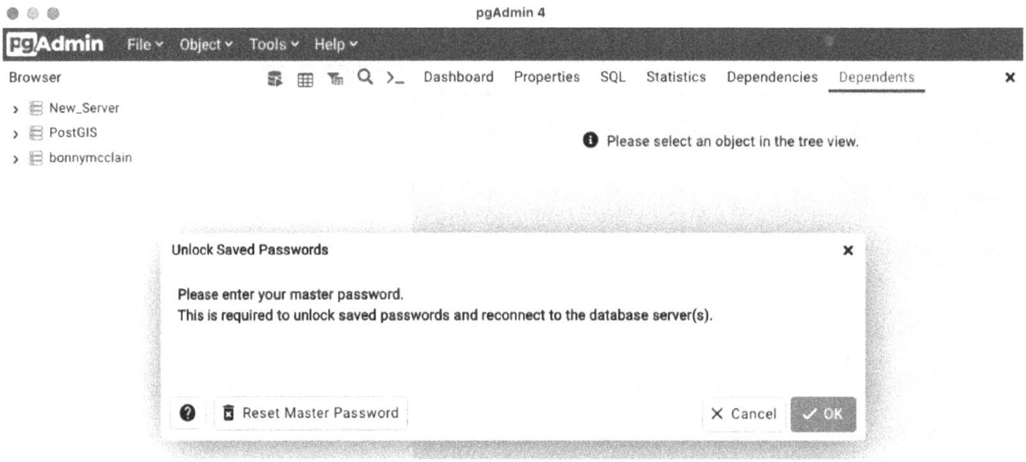

Figure 1.16 – Setting up your server in pgAdmin

In the later chapters, you will learn more about pgAdmin properties and the browser. The objective here is to install and create a working environment for exploring geospatial data using spatial SQL. Here are the steps to achieve this:

1.  The default database, Postgres, is already loaded in *Figure 1.17*, but let's create a new one.

2.  Right-click on **Databases**, and you will see the option to create a new database. *Figure 1.17* shows the steps for adding a database. You are able to customize, but for now, this is introductory for familiarizing you with options. When selecting a longer name, you will need to use underscores between words—for example, `long_database_name`:

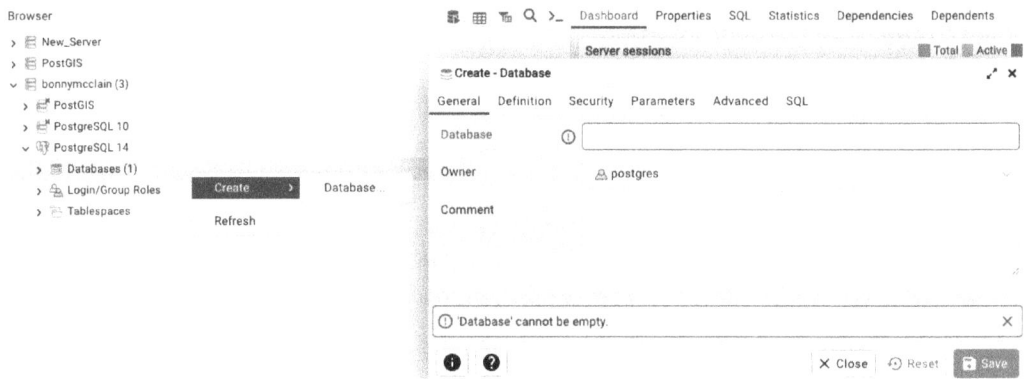

Figure 1.17 – Creating a new database in pgAdmin

By default, the new database defaults to a public schema, as shown in *Figure 1.18*. Schemas can be customized, as you will see. The tables and functions folders are also empty. The PostGIS extension has not been added yet and will need to be integrated with each database to introduce spatial functionality:

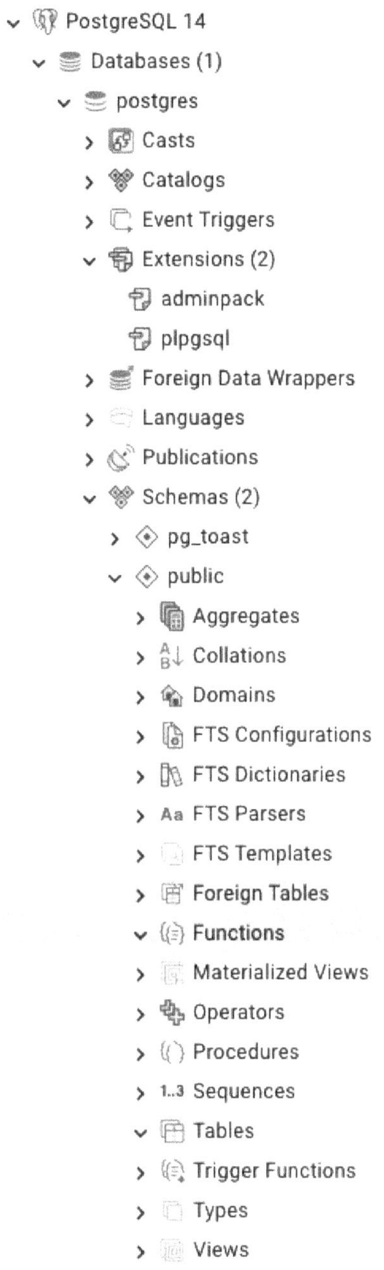

```
∨ (P) PostgreSQL 14
    ∨ ≋ Databases (1)
        ∨ ≋ postgres
            > 🔲 Casts
            > 🏵 Catalogs
            > ⊏ Event Triggers
            ∨ 🔁 Extensions (2)
                  🗐 adminpack
                  🗐 plpgsql
            > ≋ Foreign Data Wrappers
            > ⌯ Languages
            > ⌖ Publications
            ∨ 🏵 Schemas (2)
                > ◈ pg_toast
                ∨ ◈ public
                    > 🔳 Aggregates
                    > ᴬ↓ Collations
                      ᴮ
                    > 🏠 Domains
                    > 🔲 FTS Configurations
                    > 🔖 FTS Dictionaries
                    > Aa FTS Parsers
                    > ▢ FTS Templates
                    > 🔲 Foreign Tables
                    ∨ ⦅≣⦆ Functions
                    > 🔲 Materialized Views
                    > 🔁 Operators
                    > ⦃⦄ Procedures
                    > 1..3 Sequences
                    ∨ 🔲 Tables
                    > ⦅≣⦆ Trigger Functions
                    > ▢ Types
                    > 🔲 Views
```

Figure 1.18 – Customizing databases in pgAdmin

The PostGIS extension will need to be added to each database you create.

## CREATE statements

Let's go through the following set of instructions:

1. Select the **Extensions** option, right-click, and then write `CREATE EXTENSION postgis` into the query tool, as shown in *Figure 1.19*.

2. Click the arrow and run the cell. You will see the extension added on the left once you click **Extensions** and select **Refresh**.

3. Also, add `CREATE EXTENSION postgis_raster` as we will need this extension when we work with raster models. Be certain to refresh after running the code:

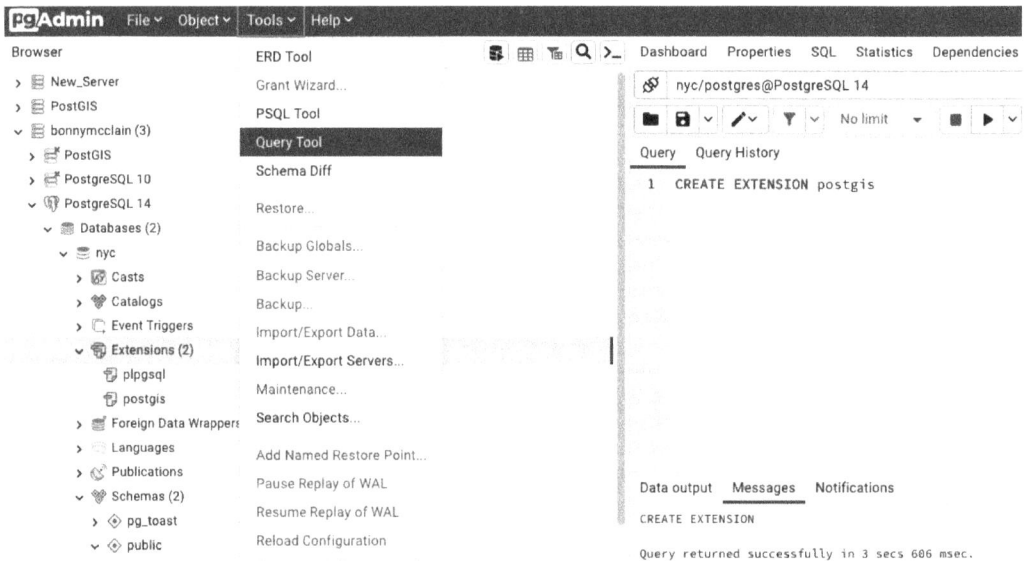

Figure 1.19 – Adding extensions to the database

4. Returning to the **Functions** option, we now see `1171` functions have been added. *Figure 1.20* shows a few of the new spatial functions updated once the `PostGIS` spatial extension has been added to the database. You will also observe a `spatial-ref_sys` table has been added. Explore the columns and note the available SRIDs within the database:

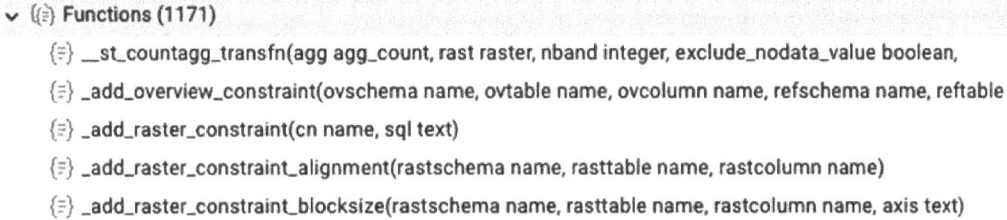

Figure 1.20 – Spatial functions have been populated in the database

You are now ready to begin working with spatial SQL in the Postgres database.

## QGIS

QGIS is an open source GIS. Although you will be able to view maps on the pgAdmin dashboard, the advanced capabilities of GIS viewing of maps, ease of data capture, and ability to integrate with Postgres will be explored. Download QGIS to your desktop to complete the installation of tools by using the following link: `https://qgis.org/en/site/index.html`. *Figure 1.21* is the same map created earlier in the graphical interface in pgAdmin rendered in QGIS.

Figure 1.21 – Downloading QGIS

Installing QGIS is straightforward, and we will customize it as we go. It is often recommended to look for the **long-term-release** (**LTR**) version as it has the most stability.

*Figure 1.22* demonstrates the additional viewing capabilities when QGIS renders the output of the same query and the same data.

Figure 1.22 – QGIS map showing the location of where in Manhattan
the percentage of impervious soil < 47.32

Let's summarize the chapter now.

# Summary

In this chapter, you were introduced to geospatial fundamentals to help understand the graphical syntax of spatial data science. You discovered a few SQL basic queries and set up your workflow for the chapters that follow. Important concepts about the properties and characteristics of vector and raster data were also introduced.

In *Chapter 2, Conceptual Framework for SQL Spatial Data Science – Geometry Versus Geography*, you will begin learning about SQL by working with the database you created in this chapter.

The fundamentals of a query-based syntax will become clearer as you discover spatial functions for discovering patterns and testing hypotheses in the datasets.

# 2
# Conceptual Framework for SQL Spatial Data Science – Geometry Versus Geography

Before analyzing geospatial data, you will need to learn about the frameworks of spatial analysis and the important distinction between geometry and geography. This is the primary difference between SQL and spatial SQL and is an important consideration when bringing data together from different datasets and—potentially—different SRIDs. Think about the difference between a map of bike rental locations your hotel might provide while you are visiting from out of town. Perhaps a local graphic designer creates it to show you sketches of the nearby ice cream shop and other amenities within a few blocks of your location. How easy do you think it would be to superimpose this map with perhaps one you brought with you from a local visitors center on your drive into town?

This is the same challenge when bringing multiple datasets together. We need to have confidence in our source of "truth."

In this chapter, we will explore the challenges of bringing datasets together, learn about a few `psql` meta-commands in terminal, begin writing SQL functions, and begin to think spatially. Specifically, we will be covering the following topics:

- Creating a spatial database
- Importing data
- Fundamentals of a query-based syntax
- Analyzing spatial relationships
- Detecting patterns, anomalies, and testing hypotheses

# Technical requirements

I invite you to find your own data if you are comfortable or access the data recommended in this book's GitHub repository at: https://github.com/PacktPublishing/Geospatial-Analysis-with-SQL.

The following datasets are being used in the examples in this chapter:

- *NYC Open Data*: `https://opendata.cityofnewyork.us`

- *DOHMH Indoor Environmental Complaints*: DOHMH Indoor Environmental Complaints

- *Census Reporter*: `https://censusreporter.org/2020/`

- *Neighborhood Tabulation Areas*: 2020 **Neighborhood Tabulation Areas** (**NTAs**)—tabular (`https://opendata.cityofnewyork.us/`)

# Creating a spatial database

In *Chapter 1, Introducing the Fundamentals of Geospatial Analytics*, maybe you created a database. If you were simply following along, you will need to create a database for working through the examples in the book. As a refresher, let's create another to hold the data for this chapter. You will use this database to explore the components of working with **PostGIS** and **spatial** data.

This is how you do it:

1.  Simply right-click the database name you created and scroll to **Query Tool**. *Figure 2.1* shows the  other options available. You will explore these in later chapters, but for now, **Delete/Drop** is how you delete a database, **Refresh** will update the database (important when we add files in QGIS), and **View/Edit Data** will render a table in your console. If you simply want to see the column headings, a quick look at **Properties** is helpful:

Count Rows

| Create | > | Table... |

Delete/Drop

Refresh...

Restore...

Backup...

Drop Cascade

Import/Export Data...

Reset Statistics

Maintenance...

Scripts          >

Truncate          >

View/Edit Data          >

Search Objects...

PSQL Tool

Query Tool

Properties...

Column...

Index...

Rule...

Trigger...

RLS Policy...

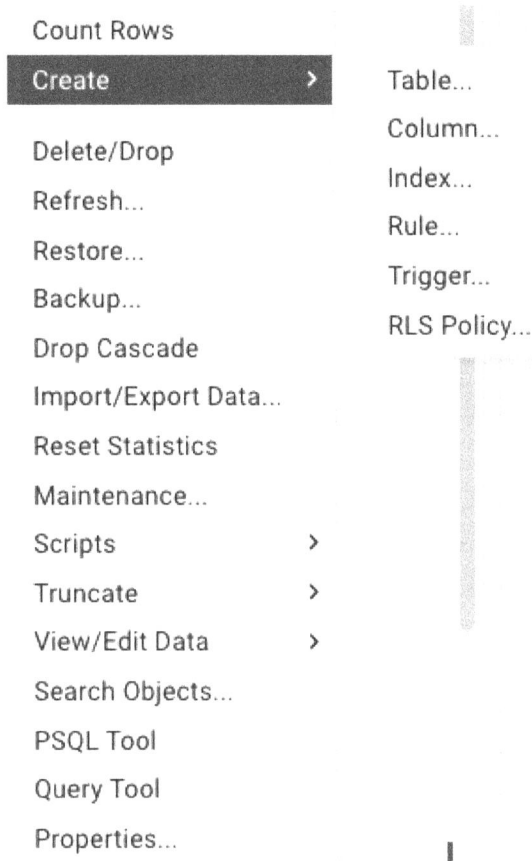

Figure 2.1 – Database options in pgAdmin

2.  Navigate to your **Browser** panel and right-click on **Databases**. You can see the databases listed on the left side of *Figure 2.2*. Here, you will name the database and select **Save**. You can create as many as you would like, but for our purposes, one will suffice.

    Now, you are able to create a database. Perhaps you've already accessed pre-existing databases. If this is the case, you will be including the server address as the host when logging in. Remember, you must enable PostGIS in any database requiring spatial functions. This is achieved with the following command:

    ```
    CREATE EXTENSION postgis
    ```

    You write this code into **Query Tool**, available in each database you create to access spatial functions.

3.  Run the code within the database you are creating. Installing the extension into a schema will help to keep the functions listed in their own schema (or container), and there will be less information you need to scroll through when working with tables and different databases.

4.   Name your database for the book or create your own hierarchy of files. I am creating a schema for the data being uploaded for each chapter. You may not require this operational level, but it works when creating files and creating folder access. One distinct advantage of creating unique schemas instead of simply relying on the public schema is accessibility. You won't need to scroll through your public schema for all of the table instances. Instead, select a specific schema and go directly to your data.

5.   Right-click on the public schema within the database you are working in. Select **Create | Schema** and add the new schema. In *Figure 2.2*, the schema has been added, and we will upload data and run queries inside the schema. The advantage for me is that each schema will hold the data for each chapter in the book.

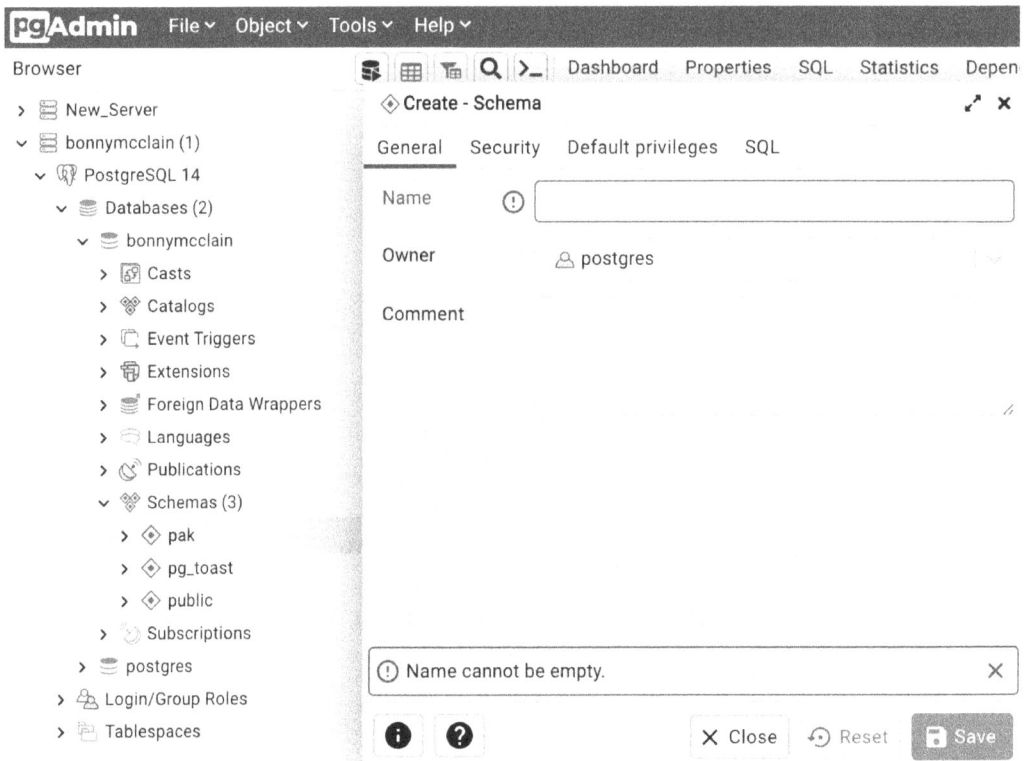

Figure 2.2 – Setting up a schema

Alternatively, you can create as many databases as you want. I have three but will be working mostly from the `bonnymcclain` database. I suggest you pick something more unique to you!

Before we import data into our new database schema, let's review a few important details about spatial functions. What is the difference between geometry and geography? Let's find out in the next section.

## Options for databases with spatial functions

In *Chapter 1, Introducing the Fundamentals of Geospatial Analytics*, you learned about the SRID. You are going to start working with different datasets and even different layers within the same dataset, and they must have the same SRS. When importing shapefiles, you noticed that checking the SRID and setting the **coordinate reference system** (**CRS**) is an important step when creating maps.

PostGIS distinguishes between geometry and geography—geometry being Cartesian for flat surfaces, and geography adding additional calculations for the curvature of the earth. Think back to working with *x*- and *y*-coordinates on a plane. If considering magnitude and direction, we now have a vector. We can also talk about vectors in three-dimensional space by including a *z*-axis. In general, if you're dealing with small areas such as a city or building, you don't need to add in the extra computing overhead for geography, but if you're trying to calculate something larger where measurements would be influenced by the earth's curved surface, such as shipping routes, for example, you need to think about geography. It would not be accurate to only consider a planar Cartesian geometry.

This is where spatial PostGIS functions can help us. The data stored in a geometry column is often a string of alphanumeric characters, known as **extended well-known binary** (**EWKB**) notation and shown in the geom column in *Figure 2.3*:

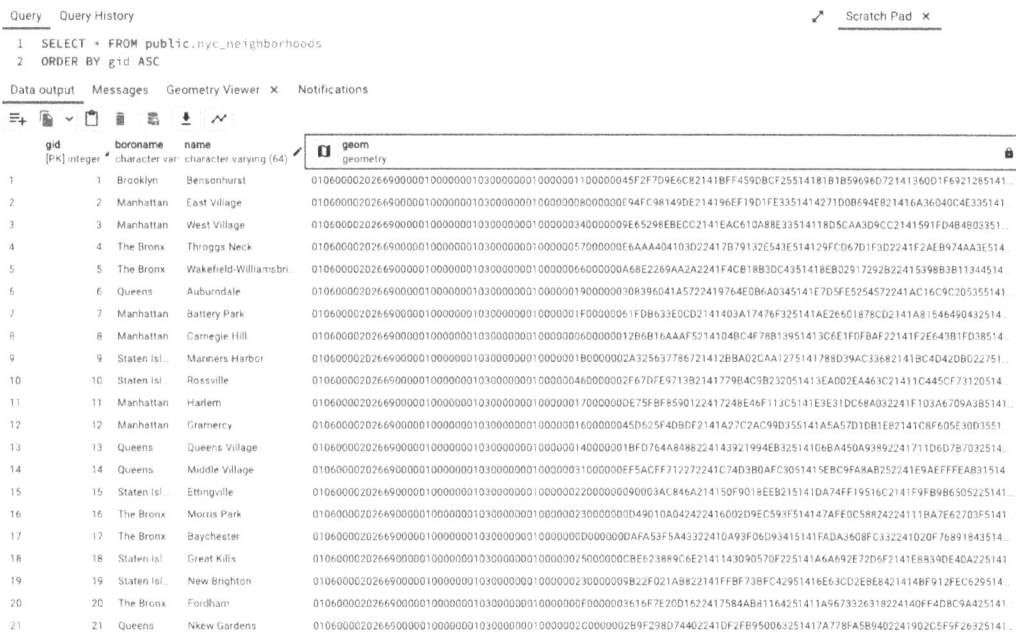

Figure 2.3 –SPACE EWKB notation for geometry

It is clear in these two instances of geometry (all the data is included on a plane, such as a map) and geography (all the data is included as points on the surface of the Earth and reported as latitude and longitude), the difference becomes relevant depending on our data questions. The location-aware `ST_AsText (geom)` function turns binary information into geometry points.

If you want to see the **geographic coordinate system (GCS)**—actual latitude longitude information—you'll need to execute `SELECT gid, boroname, name, ST_AsText(geom) FROM nyc_neighborhoods` to see the actual latitude/longitude data that's being rendered on your screen, as shown in *Figure 2.4*. The new column is now `MULTIPOLYGON`:

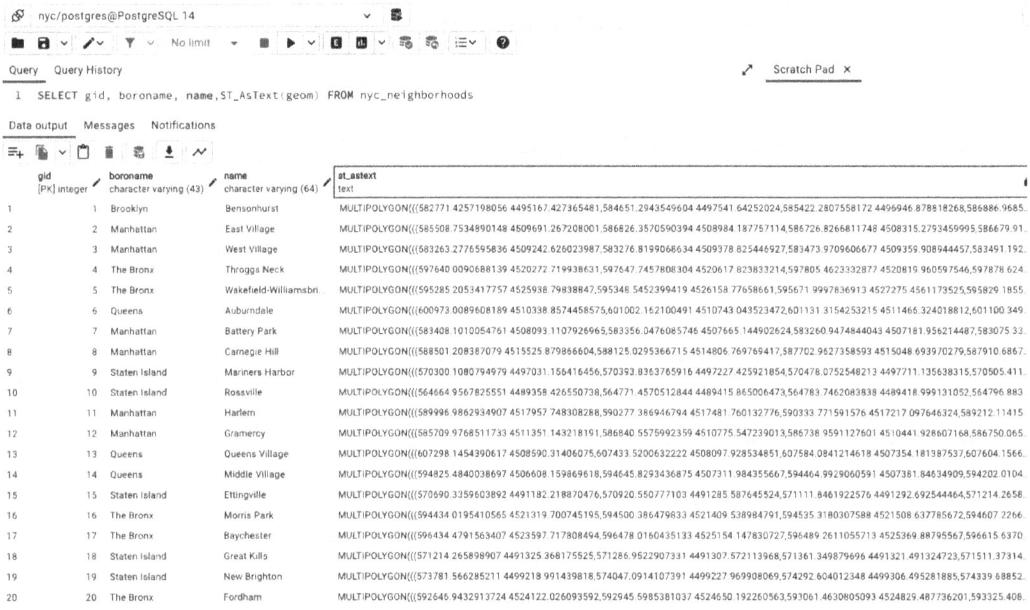

Figure 2.4 – Looking at the geometries of a single table

You will need to know the types of geometries listed in your database as well as their projections. Every database includes a `spatial_ref_sys` table and will define the SRS known for each database. This will matter a little later. There is also a `geometry_columns` table or a view that shows all of the features of the `f_table_schema` designation on the columns in your database, with the following query:

```
SELECT * FROM geometry_columns;
```

Here is how it looks:

Data output    Messages    Notifications

| | f_table_catalog<br>character varying (256) | f_table_schema<br>name | f_table_name<br>name | f_geometry_column<br>name | coord_dimension<br>integer | srid<br>integer | type<br>character varying (30) |
|---|---|---|---|---|---|---|---|
| 1 | bonnymcclain | ch2 | buildingFP | geom | 2 | 4326 | MULTIPOLYGON |
| 2 | bonnymcclain | ch2 | streetNYC | geom | 2 | 4326 | MULTILINESTRING |
| 3 | bonnymcclain | ch2 | streetrating | geom | 2 | 4326 | MULTILINESTRING |
| 4 | bonnymcclain | ch2 | buildingp | geom | 2 | 4326 | POINT |
| 5 | bonnymcclain | ch3 | DOHMH_Indo... | geom | 2 | 4326 | POINT |
| 6 | bonnymcclain | ch3 | acs2020_5yr_... | geom | 2 | 4326 | MULTIPOLYGON |
| 7 | bonnymcclain | ch3 | nyc_censustr... | geom | 2 | 4326 | MULTIPOLYGON |
| 8 | bonnymcclain | ch3 | nyc_censustr... | geom | 2 | 4326 | MULTIPOLYGON |
| 9 | bonnymcclain | ch2 | nyc_subway_... | geom | 2 | 4326 | POINT |
| 10 | bonnymcclain | ch2 | nyc_census_... | geom | 2 | 4326 | MULTIPOLYGON |
| 11 | bonnymcclain | ch2 | DOHMH | geom | 2 | 4326 | POINT |
| 12 | bonnymcclain | ch2 | nyc_streets | geom | 2 | 4326 | MULTILINESTRING |
| 13 | bonnymcclain | ch2 | nyc_neighbor... | geom | 2 | 4326 | MULTIPOLYGON |
| 14 | bonnymcclain | ch3 | occupancyst... | geom | 2 | 4326 | MULTIPOLYGON |
| 15 | bonnymcclain | ch3 | dec2020_GQ | geom | 2 | 4326 | MULTIPOLYGON |
| 16 | bonnymcclain | ch3 | census2020_... | geom | 2 | 4326 | MULTIPOLYGON |
| 17 | bonnymcclain | ch2 | census2020_... | geom | 2 | 4326 | MULTIPOLYGON |
| 18 | bonnymcclain | ch2 | dec2020_Gro... | geom | 2 | 4326 | MULTIPOLYGON |
| 19 | bonnymcclain | ch3 | acs2020_5yr_... | geom | 2 | 4326 | MULTIPOLYGON |
| 20 | bonnymcclain | ch3 | 2020_NTA | geom | 2 | 4326 | MULTIPOLYGON |
| 21 | bonnymcclain | ch2 | acs2020_5yr_... | geom | 2 | 4326 | MULTIPOLYGON |

Figure 2.5 – Display of geometry types in a single database

Certain functions need to be in a particular format. Mathematical calculations, for instance, require integer or floating-number formats. SQL CASE statements are useful in addressing mixed-use columns in SQL tables. This is the basic format for CASE summaries:

```
SELECT
    CASE
        WHEN GeometryType(geom) = 'POLYGON' THEN ST_Area(geom)
        WHEN GeometryType(geom) = 'LINESTRING' THEN ST_
Length(geom)
        ELSE NULL
    END As measure
FROM sometable;
```

When joining different tables, this will matter. If the tables are not consistent, you will get an error.

## Exploring databases with psql

Commands in a terminal are indicated by \, followed by a command and any arguments. Although each Postgres server is only able to access a single port at a time, it is possible to manage many databases. There are a few quick commands you can run for meta-commands in the server to list or switch databases on the fly. *Chapter 8, Integrating with QGIS* provides detailed instruction for creating your conda environment in terminal. Please refer there to set up. Enter the following into your terminal: conda activate sql. To find the version of psql, type the following:

```
(sql) MacBook-Pro-8:~ bonnymcclain$ psql
psql (14.4)
databasename=# \list
```

You might be curious about the template0 and template1 databases. These are used by the CREATE DATABASE command. The postgres default database is also listed in *Figure 2.6*:

```
bonnymcclain=# \list
                                     List of databases
     Name      |    Owner     | Encoding |   Collate   |    Ctype    |   Access privileges
---------------+--------------+----------+-------------+-------------+-----------------------
 bonnymcclain  | bonnymcclain | UTF8     | en_US.UTF-8 | en_US.UTF-8 |
 nyc           | postgres     | UTF8     | en_US.UTF-8 | en_US.UTF-8 |
 postgres      | postgres     | UTF8     | en_US.UTF-8 | en_US.UTF-8 |
 template0     | postgres     | UTF8     | en_US.UTF-8 | en_US.UTF-8 | =c/postgres          +
               |              |          |             |             | postgres=CTc/postgres
 template1     | postgres     | UTF8     | en_US.UTF-8 | en_US.UTF-8 | =c/postgres          +
               |              |          |             |             | postgres=CTc/postgres
(5 rows)
```

Figure 2.6 – Creating a database in terminal

To list the tables in your database, enter the \dt command.

If nothing is returned, you aren't actually connected to a database, or the database (likely the postgres default) does not have any tables.

\c is a shortcut for \connect, and it allows you to switch to a different database, as illustrated in the following code snippet. Switching to a different database lists the tables and schema for each file in a format similar to what is shown in *Figure 2.7*, depending on the number of tables:

```
\c nyc
```

The output is shown, along with the database where the tables are located:

```
                       List of relations
 Schema |                    Name                        | Type  |    Owner
--------+------------------------------------------------+-------+--------------
 public | Energy_and_Water_Data_Disclosure               | table | postgres
 public | MN_IApp                                        | table | bonnymcclain
 public | MN_LCpp                                        | table | bonnymcclain
 public | MN_pIApp                                       | table | bonnymcclain
 public | NHoodNam_Centroids                             | table | postgres
 public | geo_export_2a4bb031-f146-4f3c-b64f-1aa3a448915b | table | postgres
 public | geometries                                     | table | postgres
 public | manhattan_geopackage                           | table | bonnymcclain
 public | nyc_census_blocks                              | table | postgres
 public | nyc_homicides                                  | table | postgres
 public | nyc_neighborhoods                              | table | postgres
 public | nyc_streets                                    | table | postgres
 public | nyc_subway_stations                            | table | postgres
 public | o_128_srtm                                     | table | postgres
 public | o_16_srtm                                      | table | postgres
 public | o_256_srtm                                     | table | postgres
 public | o_2_srtm                                       | table | postgres
 public | o_32_srtm                                      | table | postgres
 public | o_4_srtm                                       | table | postgres
 public | o_64_srtm                                      | table | postgres
 public | o_8_srtm                                       | table | postgres
 public | qgis_projects                                  | table | postgres
 public | spatial_ref_sys                                | table | postgres
 public | srtm                                           | table | postgres
(24 rows)
```

Figure 2.7 – Listing the files in your database

You are also able to see the schemas connected to your database by writing the following command:

```
\dn
```

I tend to rely on terminal for troubleshooting, so these are the most useful commands for my workflow. For example, I had two versions of PostgreSQL in pgAdmin. I forgot that this is absolutely possible but they would need to have different ports. I thought my databases were connected, but checking in terminal, I realized that it was a port issue and was able to update.

> **Note**
>
> If you are interested in more advanced queries, the documentation for PostgreSQL includes a complete list of these options and commands:
>
> https://www.postgresql.org/docs/current/app-psql.html

There are many ways to import data into the database. The next section will present a popular way if you do not have the Windows environment on your computer.

# Importing data

The simplest way to import data, if you are on a Mac especially, is by using QGIS.

You should now have your database created in pgAdmin. Remember you will need to have PostgreSQL open to connect to the server before launching `pgAdmin`.

Now, open QGIS. The server port and the QGIS server port should both be `5432` (or whichever port you have assigned). Scroll to the **Browser** panel. If it isn't already docked on your canvas, select **View >> Panels** and select **Browser** and **Layer** to add them to your workspace.

*Figure 2.8* shows the **Browser** panel. On the left of the screenshot, you can scroll to your **Home** folder to see your folder hierarchy from your local computer. I download datasets to my `downloads` folder. Scroll to your downloads folder:

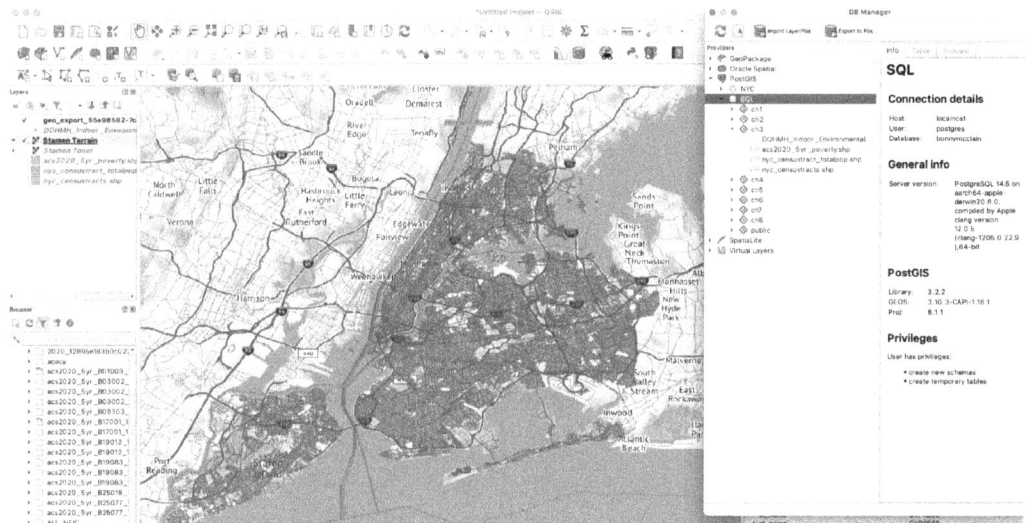

Figure 2.8 – QGIS canvas with panels and DB Manager

Right-click on the folder and open it to view a shape file or other geometry. Choose, and add `"layer to project"`. Next, I always set the CRS by right-clicking on the data layer and selecting `layer CRS`. The data is loaded in *Figure 2.8* but we have not filtered or queried it—we are simply adding the data layer to the map. Observing **DB Manager**, you can change the name of any of the files. I suggest shorter names but informative ones. You will understand why when we begin applying functions to the tables and the column names they contain. When you add the data layer to the project, you can navigate to **DB Manager** and import the shape file to your database. There is a **Database** option on the menu across the top of the console and also an icon if you have these functions included in your **toolbar** setup.

We will repeat this with different file types in later chapters, but for now, we are working with shape files. The layers in your **layers** panel are accessible here, and you can select a schema (public, or another you created) and the name of your table (self-populates with layers listed in the window)—see

*Figure 2.9*. QGIS is going to populate your **Source SRID** and **Target SRID** values but I always check **Create spatial index**. The spatial index allows queries to run faster and becomes important when working with larger datasets in PostGIS databases:

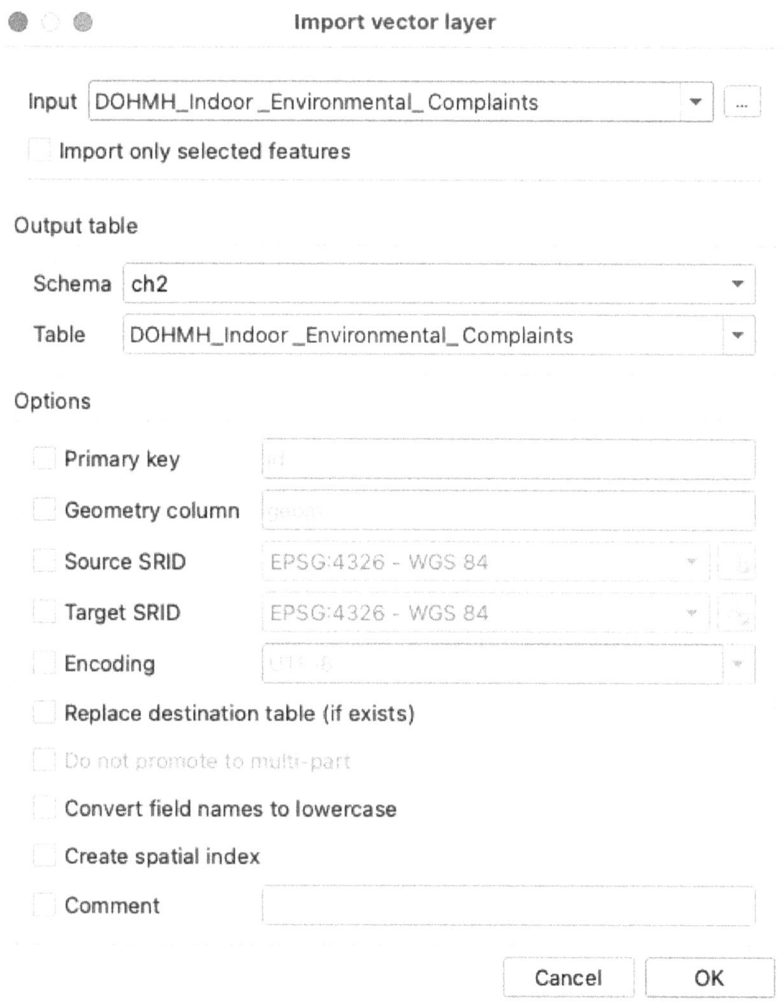

Figure 2.9 – Import vector layer in QGIS

When you are first creating a connection to PostgreSQL, you will notice the connection option in the **Browser** panel when you scroll. Right-click on the icon and create a new PostGIS connection. You can name the connection anything, but it will be listed in the **Browser** panel, so choose something relevant.

The **Host** value is localhost or the IP address of a server. These credentials should be the same as your pgAdmin login information unless you are working with multiple servers. Each server has a unique port, so make sure you have selected port 5432.

*Figure 2.10* demonstrates the PostGIS connection. **Database** is the name of the database you are connecting with and where your tables will be imported. You can select whichever name you would like for this connection. The only information that is shared with pgAdmin is the database and the data you are uploading. The other defaults are fine (we will explore more of these options in later chapters), although I suggest the **Configurations** authentication. This will keep your credentials encrypted, unlike **Basic**—although this lets you store your credentials, they will be discoverable. Perhaps this is okay if it is your personal computer, but I encourage you to habitually protect your credentials:

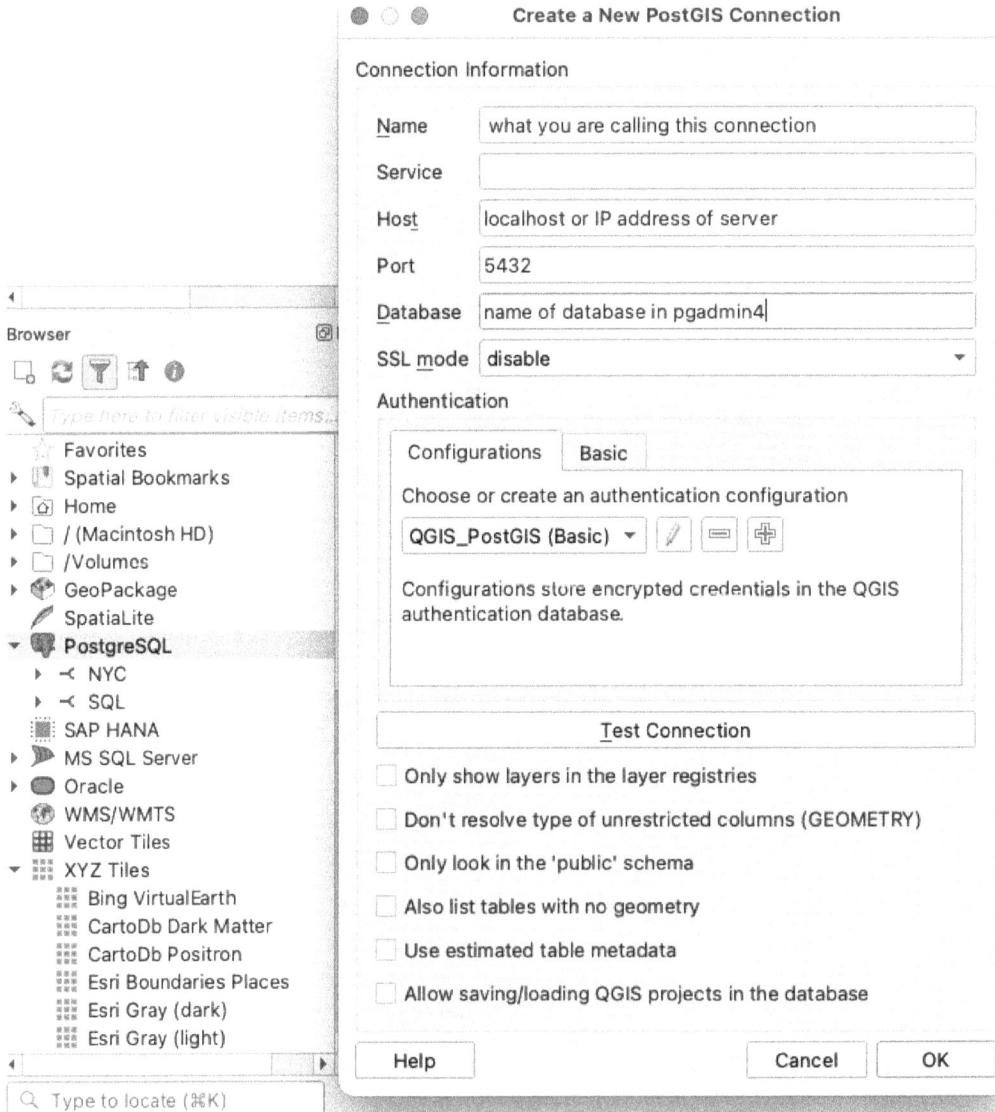

Figure 2.10 – Creating a PostGIS connection in QGIS

Now that we have some data to explore, go to `pgAdmin` and refresh the databases. The datasets you upload in QGIS will be visible in the schema under the database you accessed.

Next, let's explore the SQL syntax and how to write queries.

# Fundamentals of a query-based syntax

Don't worry if this section isn't completely intuitive. Play around with simple queries to become familiar with how the queries are executed. In a nutshell, here are the foundational elements to remember when selecting data from tables. The columns have headings that need to be entered exactly as they are spelled in the source dataset. Don't worry— we can change both table names and column headings. Have a look at the following syntax:

```
SELECT some_columns FROM some_data_source WHERE some_condition;
```

This will make sense when we begin thinking about data questions and what information to consider.

## Changing column names

I write SQL keywords such as SELECT, FROM, WHERE in all caps to improve readability, but SQL is not case-sensitive. You can decide to use lowercase if you prefer. This is relevant when you begin renaming your tables when working within a database or GIS.

The data in *Figure 2.11* is from the *Census Reporter* database (`https://censusreporter.org/2020/`) The 2020 census, although completed, does not have all of the data tables available for download, at time of publication. Here, you can select data from the *redistricting* release. We are going to download H1 or occupancy status data for **New York, NY Community Districts**. Occupancy status is an important measure of the economies of communities:

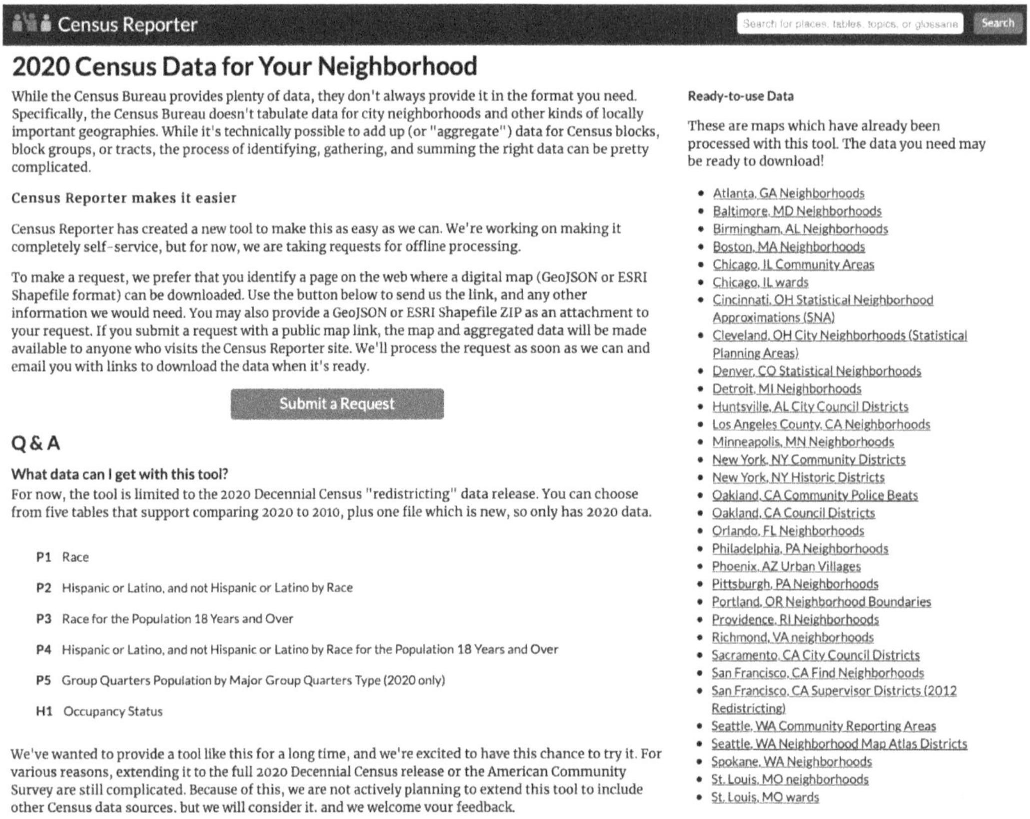

Figure 2.11 – Census Reporter 2020 census data

The folder that you downloaded contained a metadata JSON file that will explain the data you are downloading.

Repeat the following code for each column name you would like to change:

```
ALTER TABLE table_name
RENAME column_name1 TO new_column_name1;
ALTER TABLE table_name
RENAME column_name2 TO new_column_name2;
```

The metadata provides the information necessary for us to understand the data within each column. We now know what the identifiers are referring to in *Figure 2.12*:

```
{"title": "Occupancy Status", "releases": ["dec2010_pl94",
"dec2020_pl94"], "columns": {"original_id": "Geographic
Identifier", "H0010001_2020": "H1-1: Total (2020)",
```

```
"H0010002_2020": "H1-2: Occupied (2020)", "H0010003_2020":
"H1-3: Vacant (2020)", "H0010001_2010": "H1-1: Total (2010)",
"H0010002_2010": "H1-2: Occupied (2010)", "H0010003_2010":
"H1-3: Vacant (2010)", "H0010001_pct_chg": "H1-1: Total (%
change)", "H0010002_pct_chg": "H1-2: Occupied (% change)",
"H0010003_pct_chg": "H1-3: Vacant (% change)"}}
```

The column headings in our dataset are not intuitive, so we need to change them! Notice that when referring to the occupancystatus_compare_2020_2010_H1 table, I have included the schema. This isn't necessary if you rely on the public schema, but when you create them, use them. In future chapters, I will be shortening the table names. I kept them long here so that you can easily identify the table and which dataset contains the data. Recall that * in *Figure 2.12* will bring all columns into the view, but the rows are limited to the first 100:

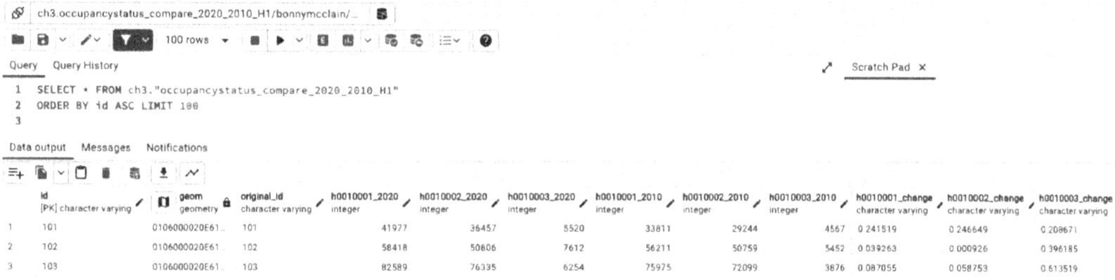

Figure 2.12 – Occupancy status of housing from the 2020 census

Notice the quotations included in the table call. There are reasons for using and not using quotations. In the early coding examples, I have included a variety of habits to bring them to your attention. There is nothing wrong with quotations, but you really only need them if you want to preserve the capitalization of the table. Postgres will convert names into lowercase unless you wrap them in quotes. More importantly, if you are using reserved words, you will need to use quotes. It is better to simply avoid using them. You will also see AS required for some. A few of the most common (USER, WHEN, TO, TABLE, CREATE, and SELECT) and a more exhaustive list can be found here: https://www.postgresql.org/docs/current/sql-keywords-appendix.html.

The following code snippet shows how we are renaming the columns to better reflect what they are -- occupied status, and vacancy status in 2010 and 2020:

```
ALTER TABLE ch3."occupancystatus_compare_2020_2010_H1"
RENAME "h0010001_2020" TO total_2020;

ALTER TABLE ch3."occupancystatus_compare_2020_2010_H1"
RENAME "h0010002_2020" TO occupied_2020;
ALTER TABLE ch3."occupancystatus_compare_2020_2010_H1"
RENAME "h0010003_2020" TO vacant_2020;
```

```
ALTER TABLE ch3."occupancystatus_compare_2020_2010_H1"
RENAME  "h0010001_2010" TO total_2010;
ALTER TABLE ch3."occupancystatus_compare_2020_2010_H1"
RENAME "h0010002_2010" TO occupied_2010;
ALTER TABLE ch3."occupancystatus_compare_2020_2010_H1"
RENAME "h0010003_2010" TO vacant_2010;
```

This chapter is focused on introducing the foundational SQL syntax. *Figure 2.13* shows the results of renaming columns. This is an important skill for enhancing readability and is especially helpful when you build out more lines of SQL syntax:

| | id [PK] character varying | geom geometry | original_id character varying | total_2020 integer | occupied_2020 integer | vacant_2020 integer | total_2010 integer | occupied_2010 integer | vacant_2010 integer |
|---|---|---|---|---|---|---|---|---|---|
| 1 | 101 | 0106000020E61... | 101 | 41977 | 36457 | 5520 | 33811 | 29244 | 4567 |
| 2 | 102 | 0106000020E61... | 102 | 58418 | 50806 | 7612 | 56211 | 50759 | 5452 |
| 3 | 103 | 0106000020E61... | 103 | 82589 | 76335 | 6254 | 75975 | 72099 | 3876 |
| 4 | 104 | 0106000020E61... | 104 | 84357 | 75001 | 9356 | 69598 | 61059 | 8539 |
| 5 | 105 | 0106000020E61... | 105 | 42372 | 33264 | 9108 | 36350 | 29636 | 6714 |
| 6 | 106 | 0106000020E61... | 106 | 99322 | 87152 | 12170 | 92267 | 82313 | 9954 |
| 7 | 107 | 0106000020E61... | 107 | 126388 | 110763 | 15625 | 120694 | 109058 | 11636 |
| 8 | 108 | 0106000020E61... | 108 | 138922 | 122565 | 16357 | 136751 | 120193 | 16558 |
| 9 | 109 | 0106000020E61... | 109 | 44900 | 41325 | 3575 | 42973 | 39856 | 3117 |
| 10 | 110 | 0106000020E61... | 110 | 61629 | 57720 | 3909 | 55513 | 49670 | 5843 |
| 11 | 111 | 0106000020E61... | 111 | 54738 | 51823 | 2915 | 50226 | 47109 | 3117 |
| 12 | 112 | 0106000020E61... | 112 | 74722 | 71076 | 3646 | 72910 | 69182 | 3728 |
| 13 | 164 | 0106000020E61... | 164 | 5 | 5 | 0 | 0 | 0 | 0 |
| 14 | 201 | 0106000020E61... | 201 | 36896 | 35495 | 1401 | 31623 | 29900 | 1723 |

Figure 2.13 – Renamed occupancy status table

Organizing the table names and columns, especially when working with census data, will help your continued exploration of the dataset. Not all columns are relevant to each query, and in the next section, you will learn how to select only columns of interest.

## Identifying tables

When you have a large dataset with numerous columns, you will want to select only those columns of interest. The basic structure of a query to select columns is shown here:

```
SELECT column_name1, column_name2
FROM table_name;
```

Using the rubric for selecting columns and populating with our columns of interest, we can see the visual in *Figure 2.14* by selecting **Geometry Viewer**:

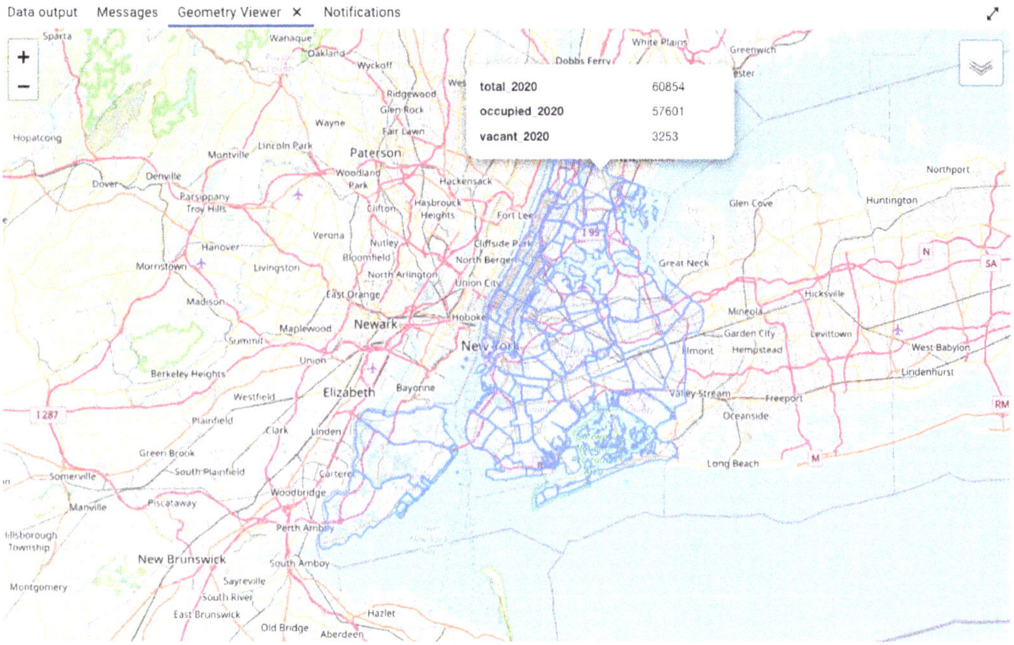

Figure 2.14 – Occupancy status

Hovering and clicking on a block assignment will also reveal data when entering the following code:

```
SELECT geom,total_2020, occupied_2020, vacant_2020
FROM ch3."occupancystatus_compare_2020_2010_H1";
```

Simple statistical questions such as the average of a particular value are easy to ask in the query builder. The answer is provided in the output:

```
SELECT AVG(occupied_2020)FROM ch3."occupancystatus_
compare_2020_2010_H1"
47462.098591549296
```

Let's get curious. *Figure 2.15* shows us where vacancy rates are between 5,520 and 9,108. To continue exploring, include the percentage change columns and rename them. It might be more accurate to know the percent change between the years instead of simply the raw number.

For now, let's see the vacancy distribution by running the following code:

```
SELECT geom,vacant_2020 FROM ch3."occupancystatus_
compare_2020_2010_H1" WHERE vacant_2020 Between 5520 AND 9108
ORDER BY id ASC
```

We can now see the block assignments with vacancy rates within our selected range:

Figure 2.15 – Filtering occupancy vacancy rates by total count

SELECT statements are a powerful tool and an important piece of syntax to locate and retrieve data.

## Introducing query-based syntax

Here is a list of some basic SQL syntax and queries. Substitute column and table names and practice. Recall that pgAdmin defaults to lowercase for table names and column headings—not a problem unless you have mixed cases. You will need to use quotations in those instances.

These basic queries are helpful to master as they provide the framework for spatial functions. Experiment with a few to see how your skills are progressing:

```
SELECT * FROM table
SELECT * FROM table LIMIT 10
SELECT column1, column2 FROM table
SELECT DISTINCT column FROM table
SELECT COUNT(DISTINCT column) FROM table
SELECT MAX(column) FROM table (if an integer)
SELECT SUM(column) FROM table (if an integer)
SELECT AVG(column) FROM table
SELECT * FROM table ORDER BY column (value)
SELECT * FROM table ORDER BY column1 ASC, column2 DESC
SELECT * FROM table WHERE column='value'
SELECT * FROM table WHERE column='value' OR column2='value'
SELECT * FROM table WHERE column='value1' AND column2>1000000
(if integer)
SELECT * FROM table WHERE column IN ('value1', 'value2',
'value3')
SELECT * FROM table WHERE column BETWEEN 1000000 AND 10000000
(if an integer)
```

The ability to translate the sample syntax and apply it to a dataset is a worthwhile endeavor. Look for datasets that include integers, and even if the output isn't a valid question, simply learning how the syntax is executed can be helpful when applied outside of a practice dataset.

## Analyzing spatial relationships

Locating datasets is a necessary step for analysis. We have uploaded several datasets from NYC Open Data. **The Department of Health and Mental Hygiene** (**DOHMH**) dataset reports environmental complaints logged by the **DOHMH**.

How can we find out where the type and frequency of these complaints are located? Are there areas where the complaints are handled more efficiently?

Let's learn how to join tables and analyze data based on additional columns and data types. Run the following code to use `INNER JOIN` on the two tables:

```
SELECT * FROM ch2."nyc_neighborhoods" JOIN ch2."DOHMH" ON name
= "NTA";
SELECT *
FROM ch2."nyc_neighborhoods",ch2."DOHMH"
WHERE name = "NTA";
```

Although tables are not visually informative in the same way as maps, they can be instrumental in formulating questions and bringing datasets together. *Figure 2.16* shows a join of tables in pgAdmin:

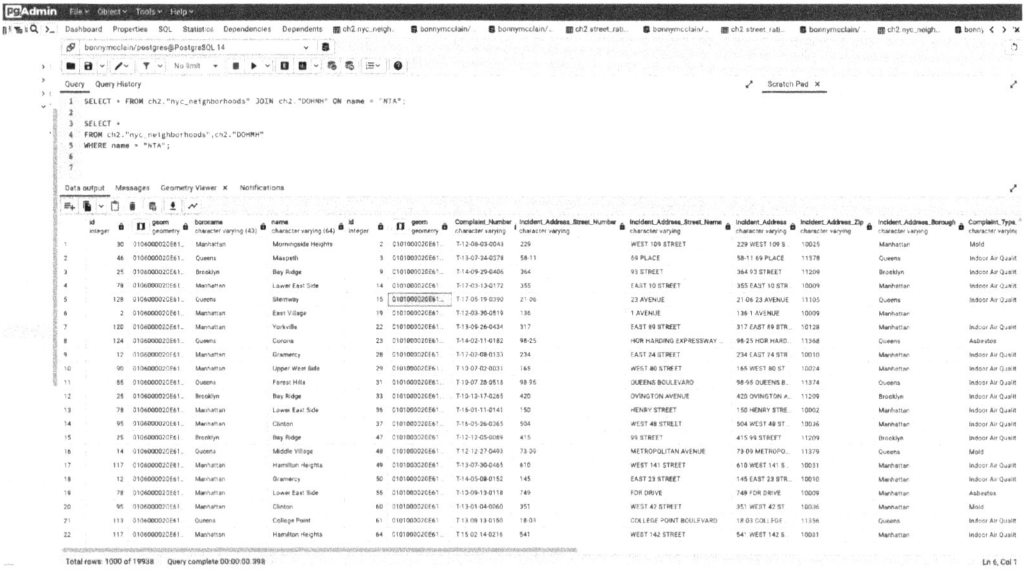

Figure 2.16 – Join of tables in pgAdmin

You can call `INNER JOIN` simply `JOIN`. You can find more information about `JOIN` available in the documentation, at `https://www.postgresql.org/docs/current/tutorial-join.html`.

Opening QGIS and adding the panel for QGIS SQL queries, we can view the data in *Figure 2.17*. Then, locate the datasets within QGIS and drag them onto the canvas:

Figure 2.17 – Running SQL queries in the QGIS console

This introduction to not only the conceptual framework of SQL but also the integration of the query-based language with the tools or platform is foundational. Spend as much time as you need here before moving on. You now have different options for building queries. We will spend more time in QGIS in later chapters now that you have been introduced to PostgreSQL and pgAdmin.

# Detecting patterns, anomalies, and testing hypotheses

Once we learn how to import data and view the tables, the next step is to ask better questions. You will eventually develop skills to build bigger queries, but in this dataset, we are now interested in complaints about indoor air quality and defining a particular neighborhood, Brownsville. The location of the complaints is displayed in *Figure 2.18*.

Run the following code in the QGIS query builder:

```
SELECT * FROM ch3."DOHMH_Indoor_Environmental_Complaints"
WHERE "Complaint_Type_311" = 'Indoor Air Quality' AND "NTA"
='Brownsville'
```

The utility of SQL queries to ask specific questions that filter data down to address the impacted communities is clearly observed:

Figure 2.18 – Filtering data in QGIS for a specific area

Historically, Brownsville was identified as the most dangerous neighborhood in Brooklyn. Additional data questions might include current rate of violence, sources of indoor air quality being problematic throughout the identified neighborhood location, type of housing, and proximity to roadways or sources of pollution. A final JOIN example unites the table geometries, returning a geometry that is the union of two input tables. *Figure 2.19* is displaying the information in **Geometry Viewer** from pgAdmin:

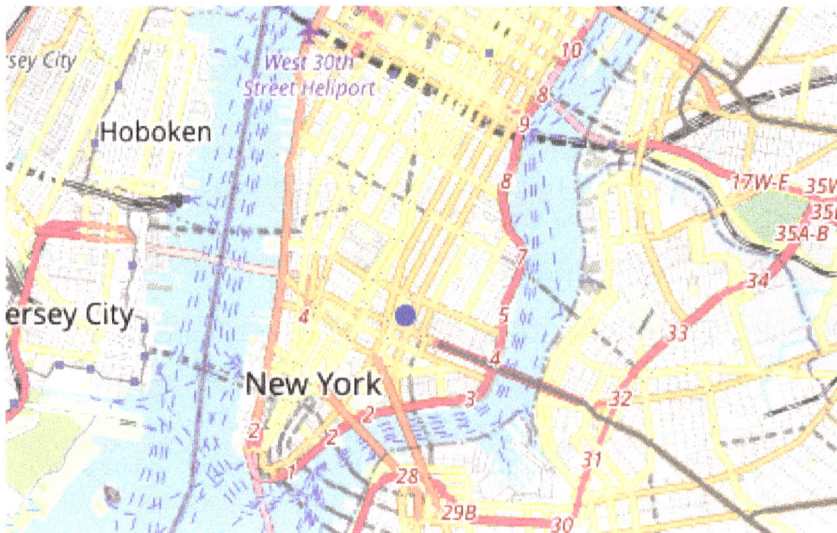

Figure 2.19 – ST_Union as single geom on JOIN

Let's continue to explore SQL expressions within the console in QGIS. I use pgAdmin as a powerful query editor to familiarize myself with data but to ask bigger questions, I move over to QGIS.

> **Note**
> *Chapter 3, Analyzing and Understanding Spatial Algorithms,* will highlight the advantages of using a GUI.

We can union the geometry so that we can locate the complaint of Mold and see where complaints are still open in the system. Locate the blue dot in *Figure 2.19* after running the following SQL query:

```
SELECT nyc_neighborhoods.geom, "DOHMH".geom,boroname, name,
"DOHMH"."Complaint_Type_311","Complaint_Status"
FROM ch2."nyc_neighborhoods" JOIN ch2."DOHMH" ON name = "NTA",
ST_Union(nyc_neighborhoods.geom,"DOHMH".geom) as singlegeom
WHERE "Complaint_Type_311" = 'Mold' AND "name"='Chinatown'
```

You now know how to create a union between two different tables in PostGIS. The aforementioned mold complaint is still an open complaint in the system. The ability to explore non-spatial information and location data has the opportunity to provide additional clues. Are there neighborhood characteristics that might influence how quickly complaints are managed? We could also look for patterns in the data to see what we might learn about the outcomes of complaint_status.

Viewing tables in pgAdmin is not the same as being able to locate a precise location on a map. Often, visualizing information brings more questions to mind for further analysis. Geospatial analysis is the tool for digging deeper.

## Summary

In this chapter, we were introduced to additional spatial considerations and how to differentiate between a Cartesian coordinate system and a spatial coordinate system. Creating and exploring databases is fundamental to building SQL skills, and both were introduced as foundational to working with datasets in the chapters that follow. We learned how to import data into QGIS that synchronizes with the database(s) we created in pgAdmin. And with this, the application of SQL fundamentals to query datasets will prepare us for problem-solving in the next chapter, using geometric data and integrating GIS functions into a modern SQL environment.

# 3

# Analyzing and Understanding Spatial Algorithms

SQL is a powerful tool not only for data exploration but also visualization and subsequent queries. Once you begin writing SQL, you should notice a pattern to writing successful queries. In this chapter, we will begin refining and expanding how we use SQL, as well as building data questions as we join data tables and learn how to run queries in QGIS as well.

The next steps include interfacing with QGIS and running geospatial queries in a GUI that will allow advanced filtering and layer styling, across databases.

This chapter will cover importing a variety of file types into databases along with connecting to databases and executing SQL queries in both `pgAdmin` and QGIS. We will also cover Spatial joins and how to visualize the output.

Following are the skills introduced:

- Developing knowledge about geographic space
- Connecting to databases and executing SQL queries
- Exploring pattern detection tools

## Technical Requirements

I invite you to find your own data if you are comfortable or access the data recommended in this book's GitHub repository at: https://github.com/PacktPublishing/Geospatial-Analysis-with-SQL.

The following datasets are being used in the examples in this chapter:

- Affordable Housing Production by Building (`https://data.cityofnewyork.us/Housing-Development/Affordable-Housing-Production-by-Building/hg8x-zxpr`)
- **Department of Health and Mental Hygiene** (**DOHMH**) Indoor Environmental Complaints

- NYC street data: `https://data.cityofnewyork.us/Transportation/Street-Pavement-Rating/2cav-chmn?category=Transportation&view_name=Street-Pavement-Rating`

- PostGIS documentation: `https://postgis.net/workshops/postgis-intro/`

- PostgreSQL documentation: `https://www.postgresql.org/docs/current/sql-createindex.html`

# Developing knowledge about geographic space

The PostGIS documentation is an important tool for improving your understanding of spatial algorithms and how to execute efficient SQL queries. As we move past inquiry and into efficiency, it is important to pay attention to additional details. These will be explained as we move through the chapter.

Let's begin by understanding **spatial indexing**. Unique to spatial databases, indexing is necessary to expedite searches and improve the speed of our queries.

But wait – you might be thinking – how do you index geometries? Isn't that what makes a spatial database different? The simple answer is a spatial index is looking at bounding boxes—not simply the lines generated from their edges. Bounding boxes are rectangular polygons containing an object or an area of interest. Identified by *xmax*, *xmin*, *ymax*, and *ymin*, a bounding box is slightly different from an extent, as it can contain an extent but does not have to be the same. The *xmin* and *ymin* coordinates describe the top-left corner of the bounding box or rectangle while *xmax* and *ymax* are the coordinates of the bottom-right corner of the bounding box or rectangle.

You can refer to the following figure:

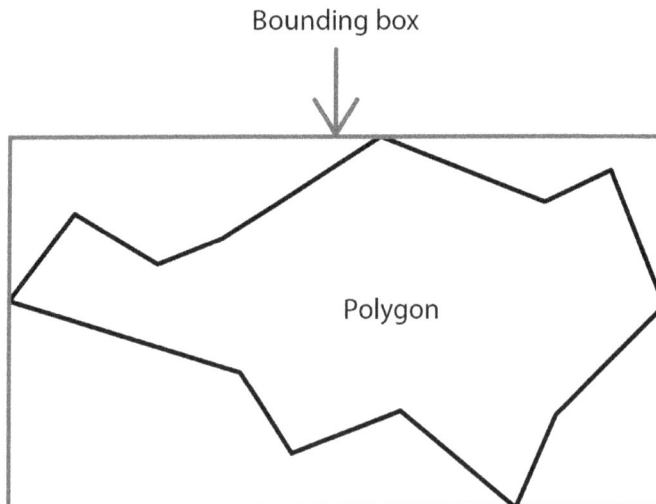

Figure 3.1 – A bounding box and an extent

When we write functions about place and location it matters how efficient (and quick) the data processes are at generating their results, especially when analyzing large datasets. The following figure is a simplified graphic showing how intersecting lines generate bounding boxes and how this expedites processing.

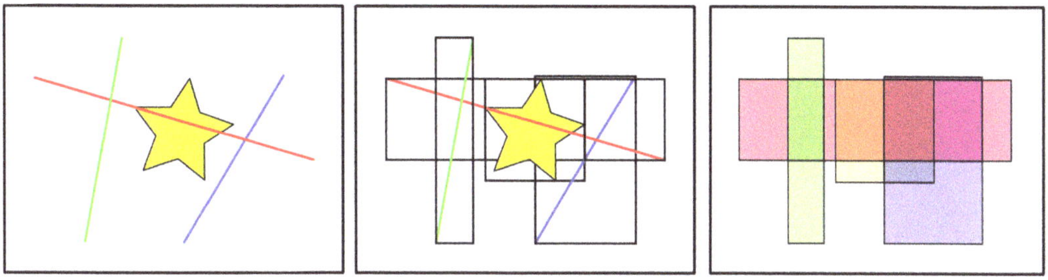

Figure 3.2 – Illustration of bounding boxes and spatial indexing

Let's look at a few spatially indexed functions.

## Connecting to databases and executing SQL queries

There is an ST_ spatial prefix, which stands for **Spatial Type**, on all evolving PostGIS functions.

This evolution is aligned with the ISO standard SQL-MM defining the spatial type and associated routines. In this case, **MM** stands for **multimedia**.

This is all you need to know about it because all prior formats have been deprecated (and are tolerated but not supported). Not all of the functions will automatically spatially index but fortunately for us, the most popular ones do. The && operator in *Figure 3.3* selects bounding boxes that overlap or touch. The operator is an index-only query.

Now run the following code and notice how many rows are returned. The output for *Figure 3.3* yields 2,790 rows that meet the criteria of having extremely low-income units.

```
SELECT * FROM ch3."Affordable_Housing_Production_by_Building"
borough
JOIN ch3."DOHMH_Indoor_Environmental_Complaints" Incident_
Address_Borough
ON borough.geom && Incident_Address_Borough.geom
WHERE 'extremely low income units' NOT LIKE '%0%'
```

Figure 3.3 – Index-only query with the && operator

Now, what happens when we add the `ST_Intersects` function? The query in *Figure 3.4* applies the `ST_Intersects` function so instead of selecting each bounding box that is intersecting with another bounding box, the more accurate function sums up boroughs that are actually intersecting the borough where the complaints are being generated:

```
SELECT * FROM ch3."Affordable_Housing_Production_by_Building"
borough
JOIN ch3."DOHMH_Indoor_Environmental_Complaints" Incident_
Address_Borough
ON ST_Intersects (borough.geom,Incident_Address_Borough.geom)
WHERE 'extremely low income units' NOT LIKE '%0%'
```

When you run the function, you will notice in *Figure 3.4* that only `2769` rows are returned. This may not be evident visually when working with a smaller set of data but this matters when we begin working with more complex data questions.

Data output    Messages    Geometry Viewer  ×    Notifications

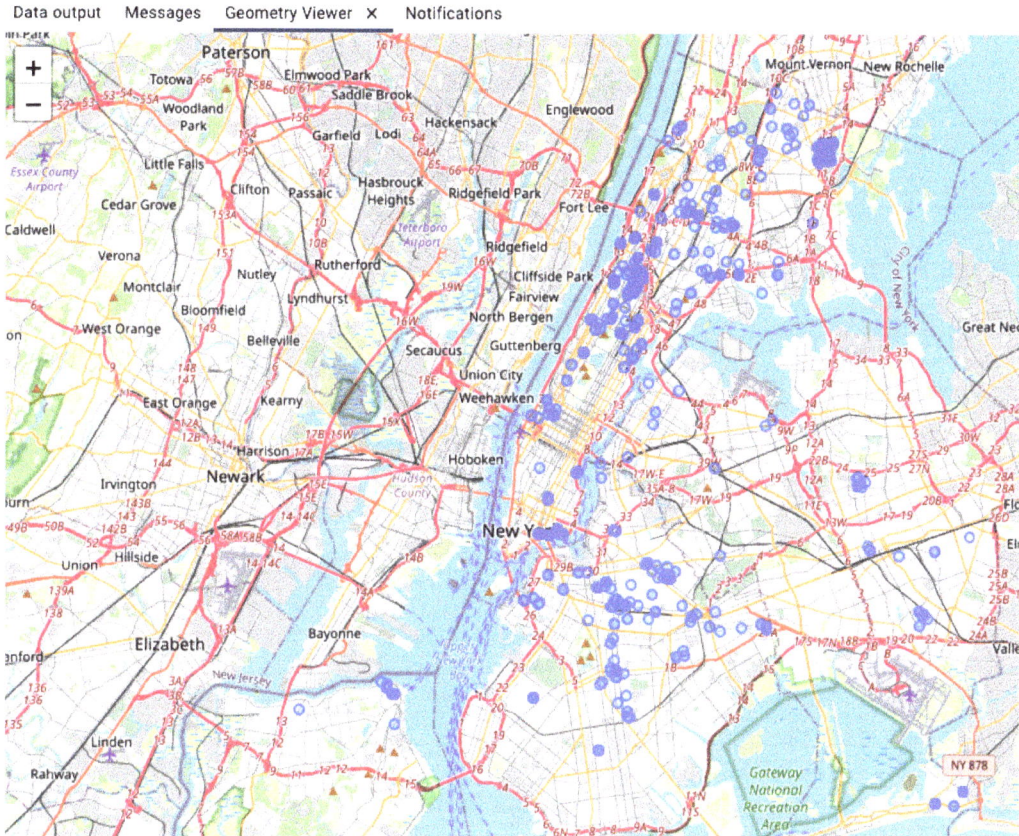

Figure 3.4 – ST_Intersect applied to identify extremely low-income units

The ANALYZE and VACUUM functions can also be run individually or together against a database (I don't recommend it, as it takes forever), a table, or even a column. The work case for me is when I have been manipulating the data by deleting data and inserting data into a table.

PostgreSQL doesn't actually remove a deleted row from the table but simply masks the row so queries don't return that row. The VACUUM function marks these spaces as reusable.

The ANALYZE function collects statistics based on the values in column tables. How are the values distributed? You can run the operation and find out.

Both of these functions need to be used cautiously, with additional information and cautions for VACUUM (https://www.postgresql.org/docs/current/sql-vacuum.html) and ANALYZE (https://www.postgresql.org/docs/15/sql-analyze.html) in the user documentation for PostgreSQL.

It is good practice to reclaim free space by running the following code snippet:

```
Analyze ch3."Affordable_Housing_Production_by_Building"
```

This is the output:

**Query returned successfully in 897 msec**

What happens when we apply the VACUUM function? The query is returned even faster:

```
Vacuum ch3."Affordable_Housing_Production_by_Building"
```

Here is the output:

**Query returned successfully in 129 msec**

I think it is easier to appreciate the SQL functions by working with real data. We are going to start building stories. But first, we need to understand the dataset.

The dataset introduced earlier is from NYC OpenData. You explored the DMOH data in *Chapter 2, Conceptual Framework for SQL Spatial Data Science – Geometry Versus Geography*. The *Affordable Housing Production* by Project Data is building level data counted towards *Housing our Neighbors: A Blueprint for Housing and Homelessness*, https://www1.nyc.gov/assets/home/downloads/pdf/office-of-the-mayor/2022/Housing-Blueprint.pdf. Estimates show that one-third of New York City residents spend 50 percent of their income on rent, and the number of children sleeping in shelters is increasing. Bold changes are being enacted and we have data to explore how this impacts the community infrastructure and how people are living.

The first step for data analysis is not only exploring the dataset but also thinking about the types of questions to explore. Interesting variables to explore include the types of units being built and how they are contributing to improving the housing and affordability crisis. For one, the housing inventory is not increasing at the same rate as the rapid population and job growth in NYC over the last few decades. The data in *Figure 3.5*, available at the provided link, allows you to explore the variables. Exploring the characteristics of a dataset is instrumental for generating SQL queries and formulating questions for deeper insights.

| project_id | text | Project ID |
| project_name | text | Project Name |
| project_start_date | floating_timestamp | Project Start Date |
| project_completion_date | floating_timestamp | Project Completion Date |
| building_id | number | Building ID |
| house_number | text | Number |
| street_name | text | Street |
| borough | text | Borough |
| postcode | number | Postcode |
| bbl | number | BBL |
| bin | number | BIN |
| community_board | text | Community Board |
| council_district | number | Council District |
| census_tract | text | Census Tract |
| neighborhood_tabulation_area | text | NTA - Neighborhood Tabulation Area |
| latitude | number | Latitude |
| longitude | number | Longitude |
| latitude_internal | number | Latitude (Internal) |
| longitude_internal | number | Longitude (Internal) |
| building_completion_date | floating_timestamp | Building Completion Date |
| reporting_construction_type | text | Reporting Construction Type |
| extended_affordability_status | text | Extended Affordability Only |
| prevailing_wage_status | text | Prevailing Wage Status |
| extremely_low_income_units | number | Extremely Low Income Units |
| very_low_income_units | number | Very Low Income Units |
| low_income_units | number | Low Income Units |
| moderate_income_units | number | Moderate Income Units |
| middle_income_units | number | Middle Income Units |
| other_income_units | number | Other Income Units |
| studio_units | number | Studio Units |
| _1_br_units | number | 1-BR Units |
| _2_br_units | number | 2-BR Units |
| _3_br_units | number | 3-BR Units |
| _4_br_units | number | 4-BR Units |
| _5_br_units | number | 5-BR Units |
| _6_br_units | number | 6-BR+ Units |
| unknown_br_units | number | Unknown-BR Units |
| counted_rental_units | number | Counted Rental Units |
| counted_homeownership_units | number | Counted Homeownership Units |
| all_counted_units | number | All Counted Units |
| total_units | number | Total Units |

Figure 3.5 – Exploring dataset characteristics

`extremely_low_income_units` are units with rents that are affordable to households earning 0 to 30% of the **area median income** (**AMI**). There is a column heading, borough, shared between other NYC datasets, so there are options to explore.

Knowing the datatype of the column provides information about the types of functions we can apply to the data.

The `extremely_low_income_units` column is of the `number` datatype.

The data is imported as a `Vector` by **Data Source Manager** in QGIS. Once you complete the fields in *Figure 3.6*, you will now have a data source with a geometry column. Often, QGIS will detect the geometry column but you may have to add the column headings (exactly as it is listed in the data) in **X_POSSIBLE_NAMES** and **Y_POSSIBLE_NAMES**:

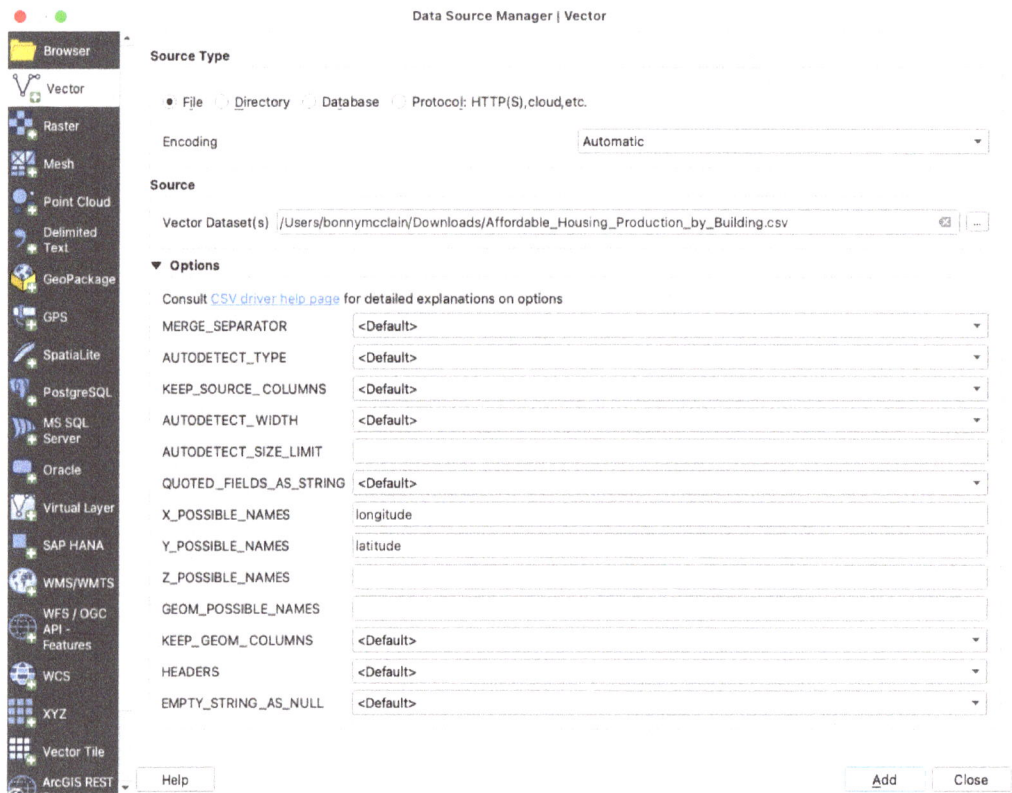

Figure 3.6 – Importing a delimited file into QGIS-defining coordinate systems in SQL

# Reasoning and key decisions about spatial concepts

Once you import data, there are a few practices that will make your analyses run smoothly. You will need consistent projections, especially when merging data from multiple datasets. When importing data, you should always check the **coordinate reference system** (**CRS**) and make sure the layers are all at the same projection. If you are exploring a specific region, for example, make sure you are using the projection for that location. You were introduced to spatial reference systems in *Chapter 1*, *Introducing the Fundamentals of Geospatial Analytics*.

QGIS will not render your CRS graphic if it is unable to confirm the projection. You can scroll through the options listed in *Figure 3.7* when you right-click on the layer:

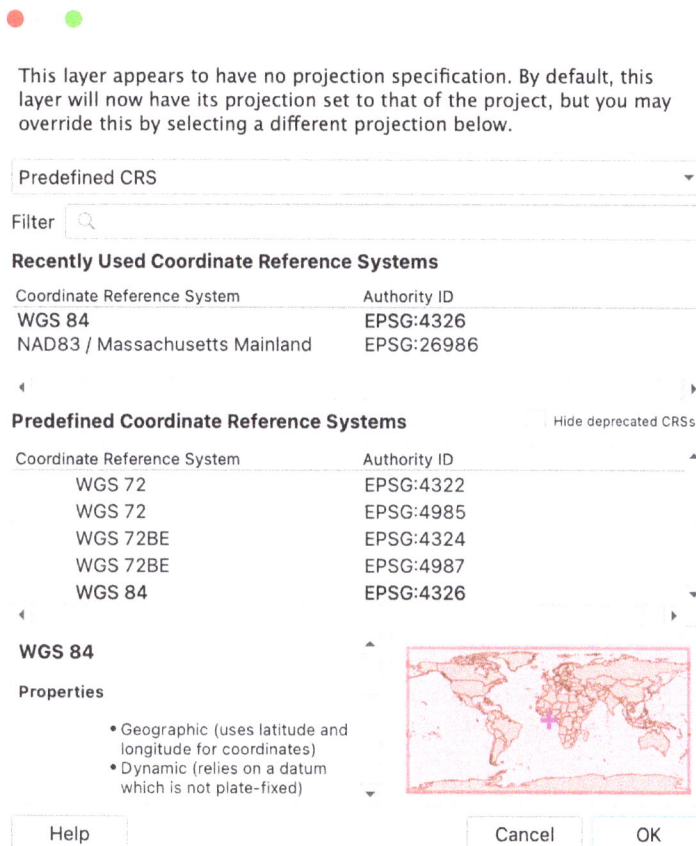

Figure 3.7 – Selecting predefined coordinate reference systems in QGIS

Once you add the layer to your project, it is easily imported into **DB Manager**. Once you import and refresh, the table you just imported will be available in pgAdmin as well. In *Figure 3.8*, the input will include the data layers you have added to your project. I typically add the layers to the project when uploading to simplify this step.

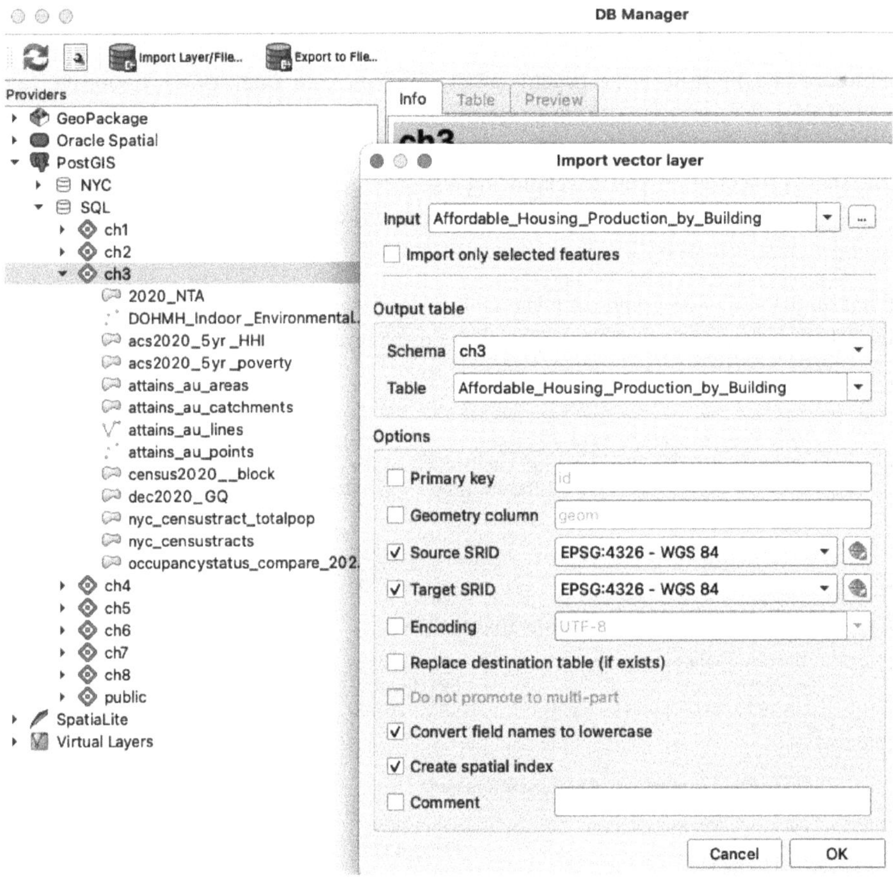

Figure 3.8 – Importing a vector layer in QGIS Database Manager

You select the **Schema** option and **Table** will display the database you selected when connecting PostGIS. Although it isn't necessary, I select **Source SRID** and **Target SRID**. This is something that QGIS already detects but for me, the act of checking the checkboxes reminds me to pay attention in case an error happens when running code.

Another habit I recommend is to check the **Convert field names to lowercase** and **Create spatial index** checkboxes. Up to this point, I simply uploaded the tables as they were listed in the datasets. This isn't a deal breaker but you may have noticed the need for quotations when referring to the **Table** instances when writing a query. Allowing field names to be lowercase allows pgAdmin to operate in the preferred environment. You will only need quotes when a reserved word for SQL is also part of the title or a number. If you wrap it in quotes, pgAdmin recognizes it as a field name and not a function.

Now that we have our data and a few best practices to move forward, let's look at specific data questions in the next section.

# Exploring pattern detection tools

There is a column in our data that needs a bit of history. It is labeled `Prevailing Wage Status` and it is derived from the Davis-Bacon Act, requiring building contractors to pay unskilled laborers the prevailing wage, which reduces opportunities for unskilled or low-skilled workers. Prevailing wages reflect the compensation paid to the majority of workers within a certain area and are often described as union wages.

Smaller firms owned by minorities are typically non-unionized and unable to pay these higher wages. The history of the act is tinged with its passage in 1931 to prevent non-unionized black and immigrant workers from working on federally funded construction projects. The modern implications of the act in NYC can be observed in *Figure 3.9*:

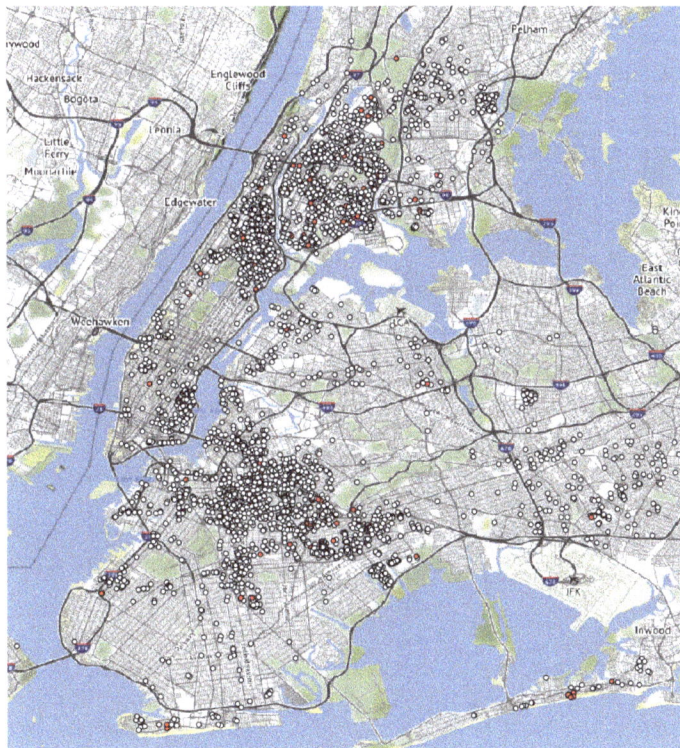

Figure 3.9 – Prevailing wages (red) and non-prevailing wages (white) in NYC

When we analyze the data, we ask questions and notice patterns in our data. After reviewing the data in pgAdmin, there is a `CONFIDENTIAL` response in the `project name` column. Since these will not be rendered in our map, we can remove them. Simultaneously, it is possible to also identify the extremely low-income units by excluding 0% values. The `NOT LIKE` operator in the code rejects values of 0 since we only want the locations with these units. The `LIKE` operator removes all zeros, `%0%`, and that is not our desired outcome. Values of 20 are fine so `0%` only removes them from the first position. *Figure 3.10* renders a map when you select the geometry column and if you hover over a point, additional attribute data is provided.

You will also learn how to simplify the syntax when working with more complex SQL strings. I want you to be familiar with what happens if we bring in our data without accommodating capitalization or if reserved words are used in column names. However, for now, by placing table names and column names in double quotes and column variables in single quotes, our code is executable:

```
SELECT * FROM ch3."Affordable_Housing_Production_by_Building"
WHERE "project name" NOT IN ('CONFIDENTIAL') AND "extremely low
income units" NOT LIKE '0%';
```

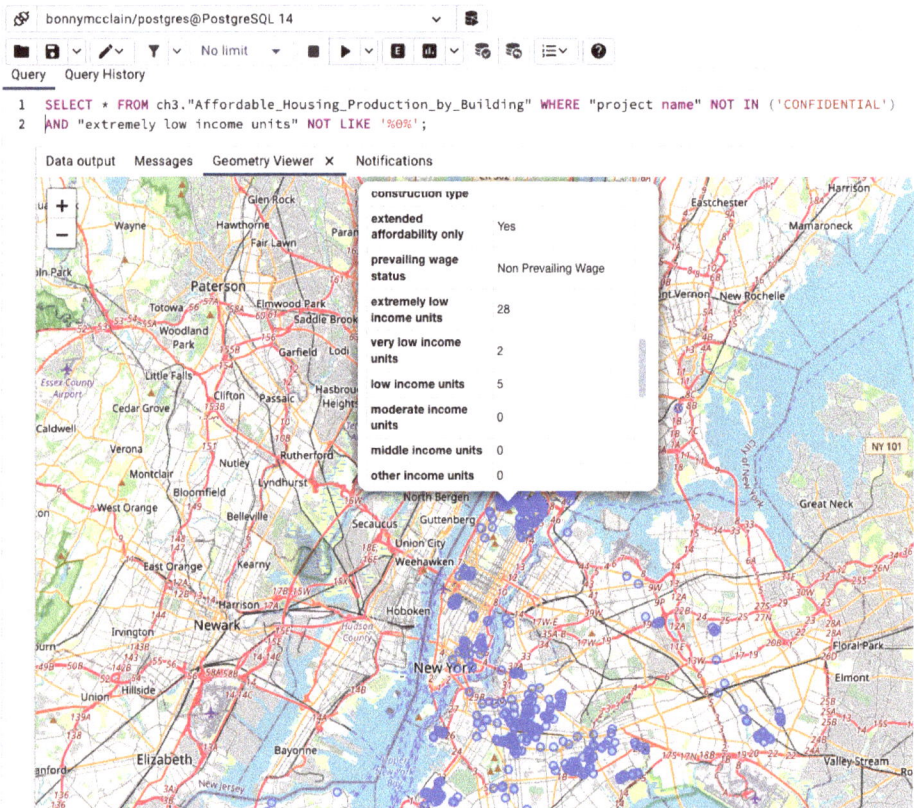

Figure 3.10 – Affordable housing data highlighting extremely low-income units

Additional exploration in areas that reflect patterns of `Extremely Low Income Units` could be the next step in your analysis.

The next chapter will introduce you to working with **census data**. The queries will be a little more complex, so let's get acquainted with a few more.

Earlier in the chapter, we talked about indexing. You can index data in a table or column. The CREATE INDEX is a PostgreSQL extension and not an SQL standard. Indexes have a default limit of 32 columns but can be expanded during the build. Indexes can also be problematic and I advise you to explore whether you indeed need one. The default index is B-tree but you can request any of the available methods, btree, hash, gist, spgist, gin, or brin.

In a spatial database, spatial queries are often between multiple tables and the geometry has to order the geometries appropriately. This can take a lot of time and an index is one way of addressing this complexity.

Run the following code if you would like to try an index. It can also be removed using DROP INDEX:

```
CREATE INDEX Affordable_Housing_Production_by_Building_idx ON
ch3."Affordable_Housing_Production_by_Building" USING GIST
(geom);
```

Now, you will be introduced to the distance operator, <->. It is basically an indexed version of ST_Distance and is used with the ORDER BY clause. The distance operator returns the two-dimensional distance between two geometries. These are called *nearest-neighbor calculations*.

In the following code, abbreviations for table names are introduced. Once the table is identified, the abbreviations are added and all functions can now refer to the abbreviations ('ah', in this example). The location included in ST_MakePoint belongs to the NYC Mayor's office. ST_MakePoint creates point geometry when assigned specified longitude and latitude values. How close is his office to the nearest New Construction projects? Not the most nail-biting question, but we can see the results following the code cell. They may not be useful since they are rendered in the units of the CRS – in this case, long/lat degrees. You can convert them into a different CRS to get unitless distances or other measures.

Select different values for EPSG 4326 in ST_SetSRID to see the different options, recalling that the spatial_ref_sys table in the public schema of spatial databases is listed in the browser of pgAdmin:

```
 ALTER TABLE ch3."Affordable_Housing_Production_by_Building"
ALTER COLUMN geom
TYPE Geometry(Point, 4326)
USING ST_Transform(geom, 4326);
SELECT ST_Distance
(ah.geom, ST_setSRID(ST_MakePoint(-74.006058,40.712772),4326))
FROM ch3."Affordable_Housing_Production_by_Building" ah
WHERE "reporting construction type" = 'New Construction'
```

```
ORDER BY ah.geom <-> ST_SetSRID(ST_Make-
Point(-74.006058,40.712772), 4326)
 LIMIT 5;
0.0030460192054607526
0.004390362171851171
0.00442200723653929
0.011724035866544517
0.012261805943660943
```

*Figure 3.11* reprojects the output reported in degrees using Albers projection (EPSG:3005) to output the distances in meters.

```
ALTER TABLE ch3."Affordable_Housing_Production_by_Building"
  ALTER COLUMN geom
  TYPE Geometry(Point, 3005)
  USING ST_Transform(geom, 3005);

SELECT ST_Distance(ah.geom, ST_setSRID(ST_MakePoint(-74.006058,40.712772),3005))
FROM ch3."Affordable_Housing_Production_by_Building" ah
WHERE "reporting construction type" = 'New Construction'
ORDER BY ah.geom <-> ST_SetSRID(ST_MakePoint(-74.006058,40.712772), 3005)
```

Data output    Messages    Notifications

| | st_distance<br>double precision 🔒 |
|---|---|
| 1 | 5200230.01199276 |
| 2 | 5202724.21833366 |
| 3 | 5202733.70377260 |
| 4 | 5203053.29803713 |
| 5 | 5203226.44923156 |

Figure 3.11 – Reprojecting from EPSG:4326 to EPSG:3005

`ST_Transform` allows you to change the projection as well, but might have a bigger impact than you intend:

```
ST_Transform(geom, 3005) as geom
from 'yourtable'
```

The following code describes a `CROSS JOIN`. During a typical `JOIN`, you can bring data together from different table names and schemas but you can also use subqueries. Remember that SQL considers `SELECT` statement outputs as a table!

The CROSS JOIN instances also are not one-to-one links to a shared variable. They are many-to-many joins based on a specific spatial condition. The distance being calculated is how far the buildings paying prevailing wages are from buildings not paying prevailing wages. The Null values are CONFIDENTIAL records that are not displayed in the data. In *Figure 3.12*, you can observe the addition of the new ST_Distance column:

```
SELECT ah1.*,
          ST_Distance(ah1.geom, ah2.geom)
     FROM (SELECT *
             FROM ch3."Affordable_Housing_Production_by_
Building"
             WHERE "prevailing wage status" = 'Prevailing
Wage') ah1
CROSS JOIN (SELECT *
             FROM ch3."Affordable_Housing_Production_by_
Building"
             WHERE "prevailing wage status" = 'Non Prevailing
Wage') ah2
```

| counted homeownership units character varying | all counted units character varying | total units character var | st_distance double precision |
|---|---|---|---|
| 0 | 118 | 118 | 0.19045979226335 |
| 0 | 154 | 154 | 0.17056891885979 |
| 0 | 31 | 31 | 0.18751121977098 |
| 0 | 22 | 22 | 0.18759373926653 |
| 0 | 32 | 32 | 0.18775696053408 |
| 0 | 94 | 94 | [null] |
| 0 | 150 | 150 | 0.17195221448414 |
| 0 | 190 | 190 | 0.04960196674528 |
| 0 | 205 | 205 | 0.17324462146918 |
| 0 | 64 | 64 | 0.03629776260322 |
| 0 | 157 | 157 | 0.00213199343338 |
| 0 | 188 | 188 | 0.20796235171299 |
| 0 | 221 | 221 | 0.18777717026571 |
| 1 | 1 | 1 | [null] |
| 0 | 78 | 78 | 0.13978009683785 |
| 0 | 164 | 164 | 0.19952251390006 |
| 0 | 103 | 103 | 0.10766409537538 |
| 1 | 1 | 1 | [null] |
| 0 | 8 | 8 | 0.11332596759788 |
| 0 | 76 | 76 | 0.11399239050480 |

Figure 3.12 – Cross joins achieve many-to-many joins and add an additional column

In the DOHMH data, we are able to view the indoor complaints data records of 311 service requests. The data is updated daily, so will differ from what you see here depending on when you download.

Let's explore the street pavement rating data from the NYC Department of Transportation. Ongoing assessments of its streets are recorded on a 1 to 10 scale as follows:

- Poor (%) rating of 1 to 3

- Fair (%) rating of 4 to 7

- Good (%) rating of 8 to 10

The final example in *Figure 3.13* features a lateral join that you can do with a single query. Every row in the `streetrating` table is now a subquery. The distance between each street rendered as POOR or FAIR is returned. Limit 1 will limit the query to return one geom per each match creating a faster query. Observations can be made regarding if there are different neighborhoods or boroughs likely to have the most streets with POOR or FAIR ratings:

```
SELECT sr1.*,
            ST_Distance(sr1.geom, sr2.geom)


   FROM (SELECT *
 FROM ch2."streetrating"
 WHERE "rating_wor" = 'POOR') sr1
 CROSS JOIN LATERAL (  SELECT geom
                           FROM ch2."streetrating"
                          WHERE "rating_wor" = 'FAIR'
                      ORDER BY sr1.geom <-> geom
                         LIMIT 1) sr2
```

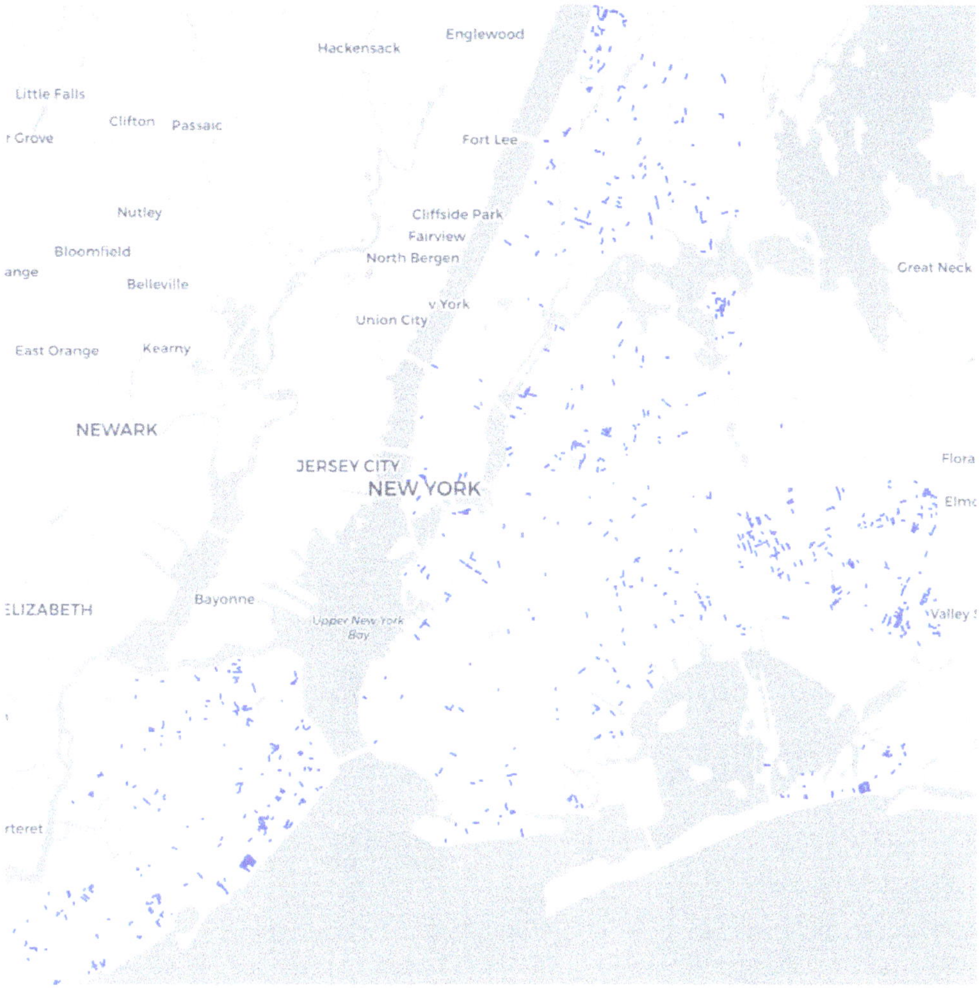

Figure 3.13 – Street ratings of POOR and FAIR in NYC using lateral joins

When you hover over the streets in pgAdmin, there is information on the length of the street, whether trucks or buses are allowed on the street, and the year of the rating. A data question might include a comparison of the different ratings over a range of years to see how different boroughs manage their street upgrades.

## Executing SQL queries in QGIS

Before now, the SQL queries have been executed in pgAdmin. If you scroll down the QGIS browser until you see PostgreSQL, as shown in *Figure 3.14*, you are now able to write your query. First, I add the layer to the project. Once you do this and update your query, the map will update.

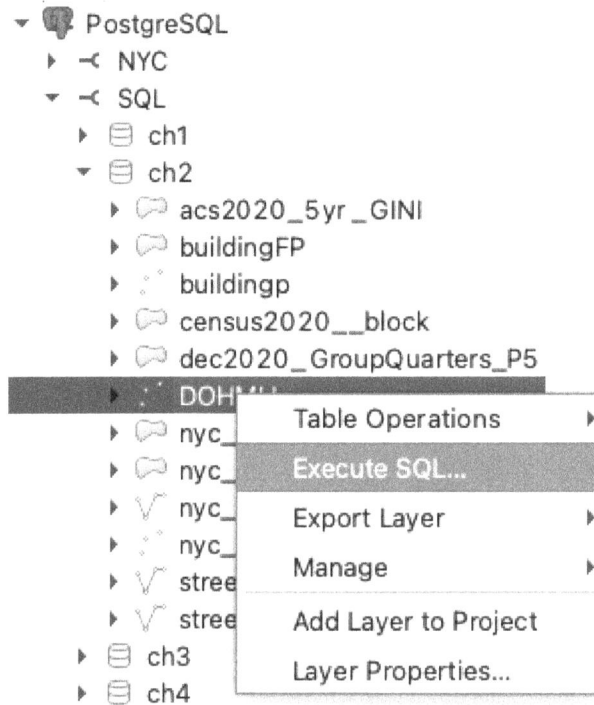

Figure 3.14 – Executing a SQL query in QGIS

The dashboard will open and you will be able to enter your query. We are looking for the results to be retrieved from the Bronx.

Execute the query and it will add the layer to your project. You will want to rename it or the default query layer label will identify the new data. The query builder in QGIS is quite advanced. In the upper-left **Layers** panel, you will see QueryLayer_Bronx. Right-click on the **Query** layer and you will see the **Update SQL** layer. The next step is to zoom in to the layer and view the changes. The lower-left panel browser in *Figure 3.15* brings you directly to your PostgreSQL tables. Don't forget to refresh when making changes so that the database updates as well.

The **Identify Results** panel docked on the right shows the data for the location I have clicked (visible as a red dot) on the console after I select the **Identify Features** icon in the top menu ribbon. If you recall, the Panels are selected by scrolling down in the **View** menu at the top of the console.  A quick assessment yields a building completed in 2021 where prevailing wages were not paid:

```
SELECT * FROM "ch3"."Affordable_Housing_Production_by_Building"
WHERE "borough" = 'Bronx'
```

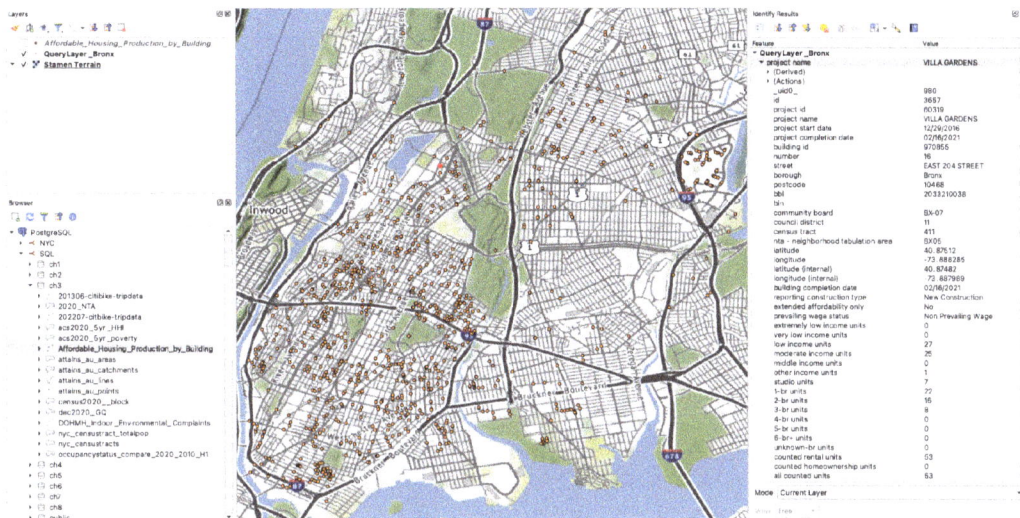

Figure 3.15 – SQL query in QGIS

You can also observe that of the 53 units, none are extremely or very low-income but the new construction dwellings do include 27 low-income units, and 25 moderate-income units. The majority are one- and two-bedroom units and all are rentals.

In later chapters, we will stray away from open NYC data but for now, these datasets are great resources to demonstrate SQL spatial functions.

# Summary

In this chapter, additional SQL expressions were presented with real-world, publicly available datasets. You learned how to import CSV files into QGIS for use across your spatial databases. Exploring different ways to join data on the fly and continue querying inside QGIS reinforced the advantages of writing queries within the graphic interface.

In the next chapter, you will be introduced to 2020 census data and continue exploring how to join tables and ask more complex questions with SQL.

# 4

# An Overview of Spatial Statistics

In this chapter, we are going to build a story by adding new spatial queries. pgAdmin is a great resource for exploring our data, executing queries, and using advanced database tools (as covered in more detail in *Chapter 7, Exploring PostGIS for Geographic Analysis*), but I usually work with QGIS when it comes to using spatial analysis and data visualization.

Census data is notoriously complex but it is worth understanding a few straightforward tasks to render the available files, as this is informative and critical to understanding the impacts of urbanization, changing population demographics, and resource scarcity (to name just a few).

In the last chapter, you were introduced to a few spatial methods such as SQL, `ST_Distance` and `ST_Intersects`. In this chapter, we will build toward discovering patterns in our data. Traditional statistical methods do not account for spatial relationships such as patterns and clusters, so we will explore data and extend analyzing data spatially.

Relying on *Census Reporter Data* from the Los Angeles 2020 package, let's see how to prepare our data and address questions as we continue learning about SQL syntax.

In this chapter, we will cover the following topics:

- Creating spatial vectors with SQL
- Working with 2020 Census data
- Running queries in QGIS
- Building prediction models

## Technical requirements

I invite you to find your own data if you are comfortable or access the data recommended in this book's GitHub repository at: `https://github.com/PacktPublishing/Geospatial-Analysis-with-SQL`.

First, here are the datasets we will access in this chapter.

> **Note**
> The files in this archive have been created from **OpenStreetMap** (**OSM**) data and are licensed under the Open Database 1.0 License.

- **Fire Hazard Severity Zones**: https://data.lacounty.gov/datasets/fire-hazard-severity-zones/explore?location=33.641879%2C-117.730373%2C7.79

- **OSM data** is refreshed daily, so the data you download may differ from what you see in the examples: http://download.geofabrik.de/north-america/us/california/socal-latest-free.shp.zip

- **Los Angeles County, CA Neighborhoods**: https://censusreporter.org/user_geo/12895e183b0c022d5a527c612ce72865/

- **Aggregate Household Income in the Past 12 Months** (in 2020 inflation-adjusted Dollars) by tenure and mortgage status: https://censusreporter.org/data/table/?table=B25120&geo_ids=05000US06037,150|05000US06037&primary_geo_id=05000US06037

- **Below Poverty (census tract) County of Los Angeles Open Data** (*Below_Poverty_(census_tract).geojson*): https://github.com/PacktPublishing/Geospatial-Analysis-with-SQL

## Creating spatial vectors with SQL

You now know how to upload your data, but what happens when things don't go as expected?

Workflows that work best also introduce you to troubleshooting guidelines and where to focus if things don't go as planned:

1. *Figure 4.1* shows the data when we select **View/Edit Data** while right-clicking on the table name in pgAdmin.

   Where is the data?

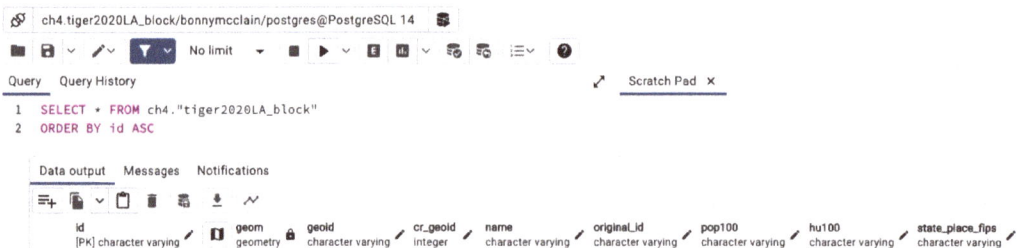

Figure 4.1 – Data not imported into pgAdmin

2.  When we head over to QGIS to import the data (as a geojson file), we are greeted by this error message:

```
Error 6
Feature write errors:
Creation error for features from #-9223372036854775808 to
#-9223372036854775808. Provider errors was:
PostGIS error while adding features: ERROR:  duplicate
key value violates unique constraint "tiger2020LA_block_
pkey"
DETAIL:  Key (id)=(athens) already exists.
Stopping after 100451 error(s)
Only 0 of 100451 features written.
```

When exploring the file uploaded from the browser (*Figure 4.2*) you may notice that there is no unique identifier. The id column contains the non-unique entry of 'athens'. Not to worry – we can add one pretty easily. I thought I would highlight this error because it is often the reason for upload failure.

| | id | geoid | cr_geoid | name | original_id | pop100 | hu100 | state_place_fips |
|---|---|---|---|---|---|---|---|---|
| 1 | athens | 0603760270... | 215 | Athens | athens | 158 | 69 | 0684116 |
| 2 | athens | 0603760270... | 215 | Athens | athens | 124 | 47 | 0684116 |
| 3 | athens | 0603760270... | 215 | Athens | athens | 102 | 47 | 0684116 |
| 4 | athens | 0603760270... | 215 | Athens | athens | 49 | 12 | 0684116 |
| 5 | athens | 0603760270... | 215 | Athens | athens | 134 | 54 | 0684116 |
| 6 | athens | 0603760270... | 215 | Athens | athens | 0 | 0 | 0684116 |
| 7 | athens | 0603760270... | 215 | Athens | athens | 47 | 13 | 0684116 |
| 8 | athens | 0603760270... | 215 | Athens | athens | 23 | 2 | 0684116 |
| 9 | athens | 0603760270... | 215 | Athens | athens | 0 | 0 | 0684116 |
| 10 | athens | 0603760280... | 215 | Athens | athens | 13 | 2 | 0684116 |
| 11 | athens | 0603760280... | 215 | Athens | athens | 32 | 5 | 0684116 |
| 12 | athens | 0603760280... | 215 | Athens | athens | 70 | 19 | 0684116 |
| 13 | athens | 0603760280... | 215 | Athens | athens | 111 | 41 | 0684116 |
| 14 | athens | 0603760280... | 215 | Athens | athens | 98 | 41 | 0684116 |
| 15 | athens | 0603760280... | 215 | Athens | athens | 196 | 56 | 0684116 |
| 16 | athens | 0603760280... | 215 | Athens | athens | 0 | 0 | 0684116 |
| 17 | athens | 0603760280... | 215 | Athens | athens | 0 | 0 | 0684116 |
| 18 | athens | 0603760280... | 215 | Athens | athens | 0 | 0 | 0684116 |
| 19 | athens | 0603760280... | 215 | Athens | athens | 0 | 0 | 0684116 |
| 20 | athens | 0603760280... | 215 | Athens | athens | 296 | 86 | 0684116 |
| 21 | athens | 0603760280... | 215 | Athens | athens | 183 | 54 | 0684116 |
| 22 | athens | 0603760280... | 215 | Athens | athens | 293 | 80 | 0684116 |
| 23 | athens | 0603760280... | 215 | Athens | athens | 250 | 75 | 0684116 |

tiger2020LA_block — Features Total: 118076, Filtered: 118076, Selected: 0

Figure 4.2 – No unique identifier in the dataset

3. You are going to add the unique identifier that was originally missing from the dataset. I think it is important to share with you how to fix this error, as the solutions floating around on the internet are far more complicated than they need to be.

   Depending on how you have set up your menu, toolbars, and panels, the icon for the **Field calculator** tool resembles a little abacus. For those of you who do not know what an abacus looks like or don't have it on your menu ribbon, access the **Processing Toolbox** window from the **Panels** menu. The **Field calculator** tool is now visible in the drop-down menu as shown in *Figure 4.3*.

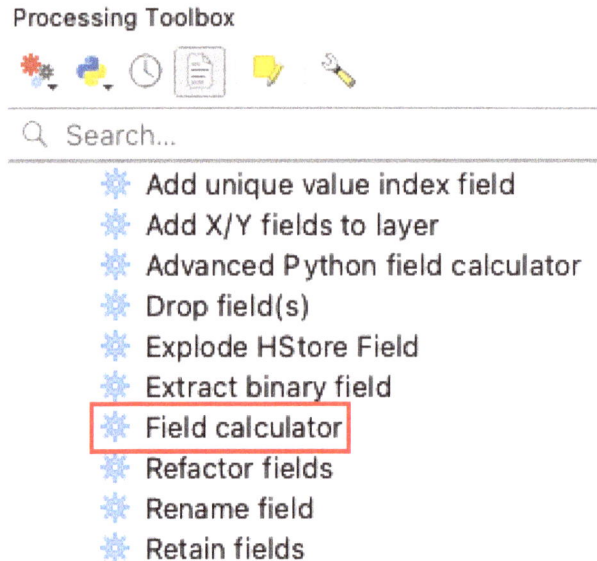

Figure 4.3 – The Field calculator tool in the Processing Toolbox window

4. When the window in *Figure 4.4* opens, enter the name of the field – here I entered an abbreviation for unique ID (`uid`) under **Output field name**, double-clicked on the **row_number** option in the center panel, and hit **OK**. This is also the window you can access to update an existing field. Notice the option in the upper-right corner of the **Rename** field panel.

Only update 0 selected feature(s)

✓ **Create a new field**                    ☐ **Update existing field**

☐ Create virtual field

Output field name  uid|

Output field type  123 Integer (32 bit)  ▼

Output field length  10  ⬍  Precision  3  ⬍

| Expression | Function Editor |

@row_number

🔍 S...  Show Help

**row_number** ▲
▸ Aggregates
▸ Arrays
▸ Color
▸ Conditionals
▸ Conversions
▸ Date and Time
▸ Fields and Val...
▸ Files and Paths
▸ Fuzzy Matching
▸ General
▸ Geometry
▸ Map Layers
▸ Maps ▼

= + - / * ^ || ( ) '\n'

Feature  ▼  ◁  ▶

Preview: 1

**variable row_number**

Stores the number of the current row.

Current value

1

ⓘ  You are editing information on this layer but the layer is currently not in edit mode. If you click OK, edit mode will automatically be turned on.

Help                              Cancel    OK

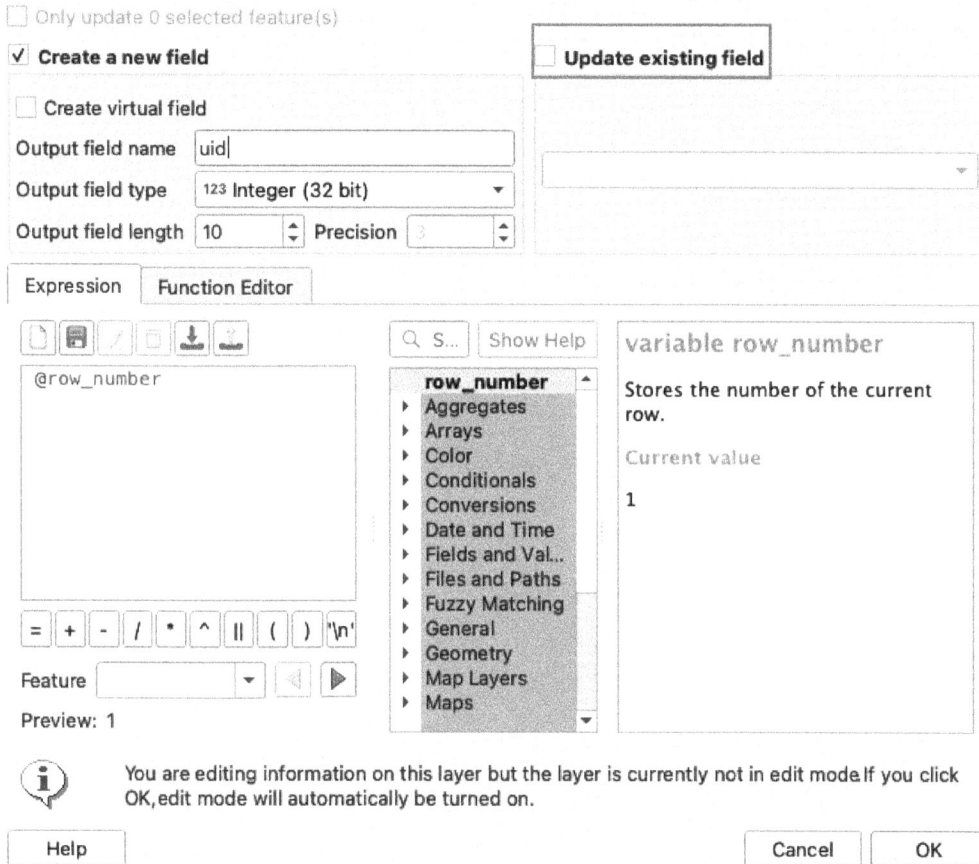

Figure 4.4 – Updating the field calculator to add another row

5.   Once you select the new unique identifier in the **Import vector layer** window in *Figure 4.5*, your table will now have a unique identifier column and you will be able to import the data. Again, it isn't necessary to check **Source SRID** and **Target SRID**, but I do so to remind myself that knowing the SRIDs can avoid problems down the road.

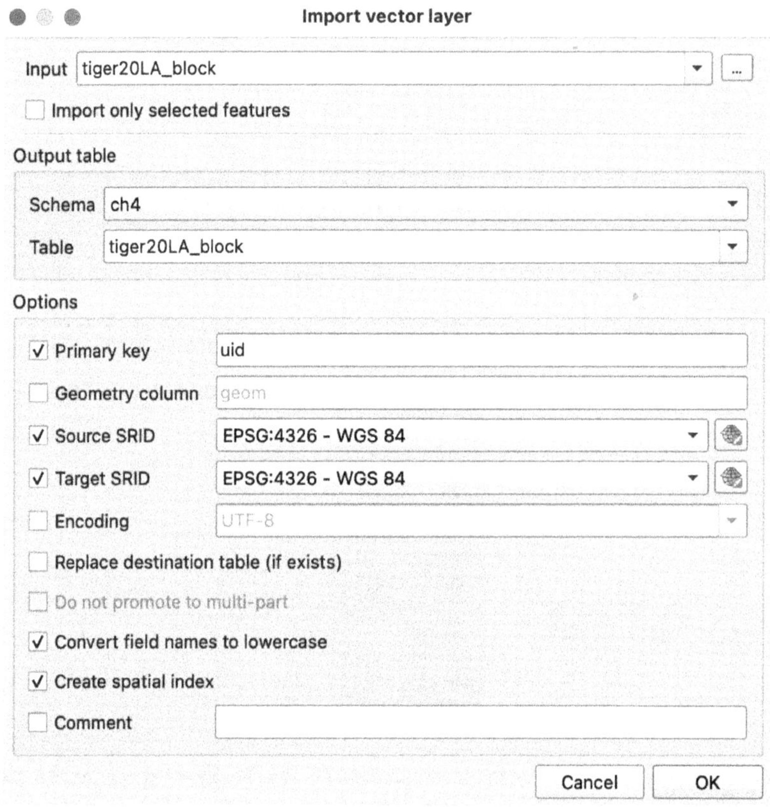


Figure 4.5 – Uploading data into QGIS and now available in pgAdmin

6. Switch over to pgAdmin and refresh the server and you will now have the unique identifier as a column heading, as in *Figure 4.6*. Census tables have column names that are not particularly intuitive. There are a variety of options to change column headings and change data types in QGIS but let's look at how this is accomplished by an SQL query.

Data output | Messages | Geometry Viewer × | Notifications

| | uid [PK] integer | geom geometry | id character varying | geoid character varying | cr_geoid integer | name character varying | original_id character varying | pop100 character varying | hu100 character varying | state_place_fips character varying |
|---|---|---|---|---|---|---|---|---|---|---|
| 1 | 1 | 0106000020E61... | athens | 060376027002000 | 215 | Athens | athens | 158 | 69 | 0684116 |
| 2 | 2 | 0106000020E61... | athens | 060376027002001 | 215 | Athens | athens | 124 | 47 | 0684116 |
| 3 | 3 | 0106000020E61... | athens | 060376027002002 | 215 | Athens | athens | 102 | 47 | 0684116 |
| 4 | 4 | 0106000020E61... | athens | 060376027002003 | 215 | Athens | athens | 49 | 12 | 0684116 |
| 5 | 5 | 0106000020E61... | athens | 060376027002004 | 215 | Athens | athens | 134 | 54 | 0684116 |
| 6 | 6 | 0106000020E61... | athens | 060376027002019 | 215 | Athens | athens | 0 | 0 | 0684116 |
| 7 | 7 | 0106000020E61... | athens | 060376027002020 | 215 | Athens | athens | 47 | 13 | 0684116 |
| 8 | 8 | 0106000020E61... | athens | 060376027002023 | 215 | Athens | athens | 23 | 2 | 0684116 |
| 9 | 9 | 0106000020E61... | athens | 060376027002028 | 215 | Athens | athens | 0 | 0 | 0684116 |
| 10 | 10 | 0106000020E61... | athens | 060376028011000 | 215 | Athens | athens | 13 | 2 | 0684116 |
| 11 | 11 | 0106000020E61... | athens | 060376028011001 | 215 | Athens | athens | 32 | 5 | 0684116 |
| 12 | 12 | 0106000020E61... | athens | 060376028011002 | 215 | Athens | athens | 70 | 19 | 0684116 |
| 13 | 13 | 0106000020E61... | athens | 060376028011003 | 215 | Athens | athens | 111 | 41 | 0684116 |

Figure 4.6 – Updating pgAdmin to display the uid column

7.  Our downloaded table depicts Hispanic and Latino populations, using `Hispanic or Latino` and `NOT Hispanic or Latino`, by race. In the downloaded folder, there is a `metadata.json` file that resembles what you can see in *Figure 4.7*. This is a key for what the column headings indicate. I paste this into the Scratch Pad inside pgAdmin, located on the right-hand side of the Dashboard.

Adding a panel to the SQL query window (or clicking on the circular arrow in the upper-left corner) creates a Scratch Pad in which you can save code snippets, or you can use it as a reference. I create a hard return after each column name description to create a list as shown in *Figure 4.8*.

● ◉ ◉                                    ⌂ metadata.json

Black or African American; American Indian and Alaska Native; Native Hawaiian and Other
Pacific Islander; Some other race (2010)", "P0040069_2010": "P4-69: White; Black or African
American; Asian; Native Hawaiian and Other Pacific Islander; Some other race (2010)",
"P0040070_2010": "P4-70: White; American Indian and Alaska Native; Asian; Native Hawaiian
and Other Pacific Islander; Some other race (2010)", "P0040071_2010": "P4-71: Black or
African American; American Indian and Alaska Native; Asian; Native Hawaiian and Other
Pacific Islander; Some other race (2010)", "P0040072_2010": "P4-72: Population of six races
(2010)", "P0040073_2010": "P4-73: White; Black or African American; American Indian and
Alaska Native; Asian; Native Hawaiian and Other Pacific Islander; Some other race (2010)",
"P0040001_pct_chg": "P4-1: Total (% change)", "P0040002_pct_chg": "P4-2: Hispanic or Latino
(% change)", "P0040003_pct_chg": "P4-3: Not Hispanic or Latino (% change)",
"P0040004_pct_chg": "P4-4: Population of one race (% change)", "P0040005_pct_chg": "P4-5:
White alone (% change)", "P0040006_pct_chg": "P4-6: Black or African American alone (%
change)", "P0040007_pct_chg": "P4-7: American Indian and Alaska Native alone (% change)",
"P0040008_pct_chg": "P4-8: Asian alone (% change)", "P0040009_pct_chg": "P4-9: Native
Hawaiian and Other Pacific Islander alone (% change)", "P0040010_pct_chg": "P4-10: Some
other race alone (% change)", "P0040011_pct_chg": "P4-11: Population of two or more races
(% change)", "P0040012_pct_chg": "P4-12: Population of two races (% change)",
"P0040013_pct_chg": "P4-13: White; Black or African American (% change)",
"P0040014_pct_chg": "P4-14: White; American Indian and Alaska Native (% change)",
"P0040015_pct_chg": "P4-15: White; Asian (% change)", "P0040016_pct_chg": "P4-16: White;
Native Hawaiian and Other Pacific Islander (% change)", "P0040017_pct_chg": "P4-17: White;
Some other race (% change)", "P0040018_pct_chg": "P4-18: Black or African American;
American Indian and Alaska Native (% change)", "P0040019_pct_chg": "P4-19: Black or African
American; Asian (% change)", "P0040020_pct_chg": "P4-20: Black or African American; Native
Hawaiian and Other Pacific Islander (% change)", "P0040021_pct_chg": "P4-21: Black or
African American; Some other race (% change)", "P0040022_pct_chg": "P4-22: American Indian
and Alaska Native; Asian (% change)", "P0040023_pct_chg": "P4-23: American Indian and
Alaska Native; Native Hawaiian and Other Pacific Islander (% change)", "P0040024_pct_chg":
"P4-24: American Indian and Alaska Native; Some other race (% change)", "P0040025_pct_chg":

Figure 4.7 – Census metadata file describing column contents

8.  Using the Scratch Pad as a reference, insert the existing table name and then the table name you would like in your table. A few things to note with this code example are as follows:

   • The schema is included since I have several different schemas for building the book chapters. If you have a single schema or are using public, this isn't needed.

- Although the `metadata.json` file lists the column titles with an uppercase P, quickly checking the **Properties** tab when right-clicking on the name of the table will show you how it is actually rendered in pgAdmin:

```
ALTER TABLE ch4."HispLat18_LA_P4"
RENAME "p0040002_2020" TO "Total (2020)"
```

You will need to match this. Again, you will notice that opting for lowercase and simplifying table names as much as feasible while also being able to identify the data will be important as we tackle more detailed queries.

# Renaming table columns – US census

In *Figure 4.8*, we are changing the column headings of the data that we will need. The ALTER TABLE function will make the requested changes and you can customize as many columns as you would like.

Figure 4.8 – Renaming the census tables for clarity

One of the most challenging aspects of working with census data is solved by the ability to create column headings that are clear and descriptive.

## What do the census table codes mean?

The data you will use is redistricting data from the 2020 decennial release. Census Reporter simplifies the task so that you can focus on learning how to write SQL queries but let me give you a quick overview.

The documentation for the decennial release can be found here: https://www2.census.gov/programs-surveys/decennial/2020/technical-documentation/complete-tech-docs/summary-file/2020Census_PL94_171Redistricting_NationalTechDoc.pdf.

The files accessed in this chapter (in bold) include *P4: Hispanic or Latino, and not Hispanic or Latino by Race for the Population 18 Years and Over*; *P5: Group Quarters Population by Major Group Quarters Type*; and *H1: Occupancy Status*:

- *P1: Race*

- *P2: Hispanic or Latino, and not Hispanic or Latino by Race*

- *P3: Race for the Population 18 Years and Over*

- *P4: Hispanic or Latino, and not Hispanic or Latino by Race for the Population 18 Years and Over*

- *P5: Group Quarters Population by Major Group Quarters Type* (2020 only)

- *H1: Occupancy Status*

The types of questions that might be of interest include demographic information such as ethnicity or race and group quarters, allowing us to look at correctional facilities or access to residential treatment centers and skilled nursing facilities. Occupancy status is also an important metric when assessing the nature of neighborhoods. Are the residential properties owner-occupied, rentals, or vacant?

QGIS has other options for renaming the column headings. Two that are straightforward include opening the **Processing Toolbox** window in **Panels** and opening **Vector table**. When you scroll down, you will see **Rename field** as shown in *Figure 4.9*.

**Layer Options** also has the option to rename the field names, or you can run the SQL query directly in QGIS.

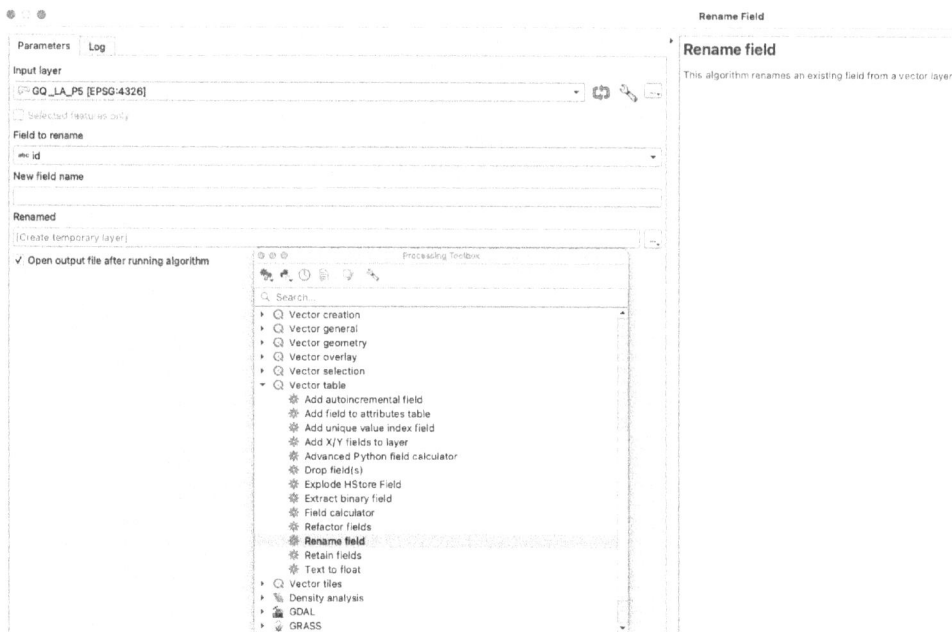

Figure 4.9 – Renaming field names in QGIS

The pgAdmin interface is useful for exploring the datasets but if you are using macOS, you will need to upload the data into QGIS first or use terminal as described earlier. One note on file types – if the option is available, GeoJSON or **Open Geospatial Consortium** (**OGC**) GeoPackage formats (`http://www.geopackage.org`) are often the preferred formats.

Yes, shape files are still the leading format thanks to the dominance of ESRI products in the marketplace and the wide support in existing software products but they come with a host of limitations. To name a few, limited support for attribute types such as arrays and images, character limits for attribute names, and the lack of a map projection definition are the ones that disrupt my workflow most frequently.

*Figure 4.10* requests specific columns by replacing * with specific columns. First, we select the renamed `Vacant(% change)` column from the table (also renamed) to `occupancy_la`. It would be difficult to detect a pattern without filtering to locations where the vacancy rate in 2020 was higher than in 2010. You can select different metrics but you will need to be certain that fields that require mathematical computation are integers or numeric.

The output shows the areas with higher vacancy rates when compared to the 2010 census. Run it without the `WHERE` clause. What do you notice?

```
SELECT "vacant_percentchange", "geom" from ch4."occupancy_la"
WHERE "vacant_2020" > "vacant_2010"
```

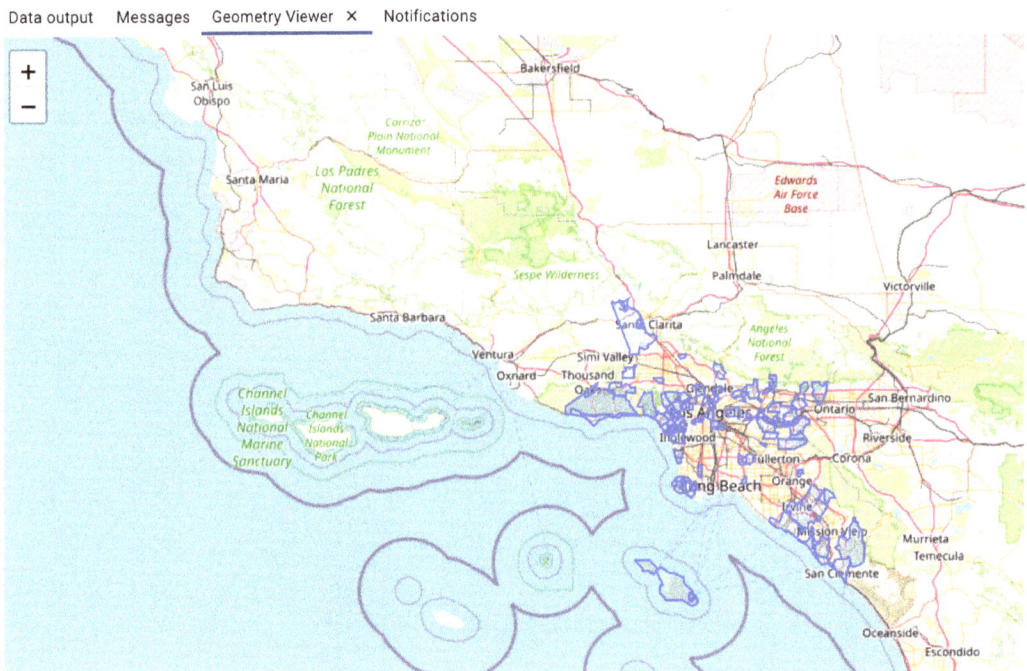

Figure 4.10 – The SQL query to select vacant residences in Los Angeles County

These scenarios are specifically for gaining familiarity with basic SQL before expanding the queries to address multi-step data questions. In the example in *Figure 4.11*, we want to see census blocks where there was a greater percentage change in Hispanic or Latino populations than white populations in the same area. We can see areas where this was not the result. More data and a deeper analysis are needed before we can think about this statistically.

Census data can be used to explore patterns of demographic shifts by examining natural changes such as births and deaths or domestic migration, occurs (due to employment opportunities and other causes unique to the pandemic that are still being explored).

Enter the following query in the query editor. This code segment compares the population growth from 2010 to 2020 of primarily Hispanic or Latino households to white alone.

```
SELECT * FROM ch4."HispLat18_LA_P4" WHERE "Hispanic or
Latino(%)"> "White alone(%)"
```

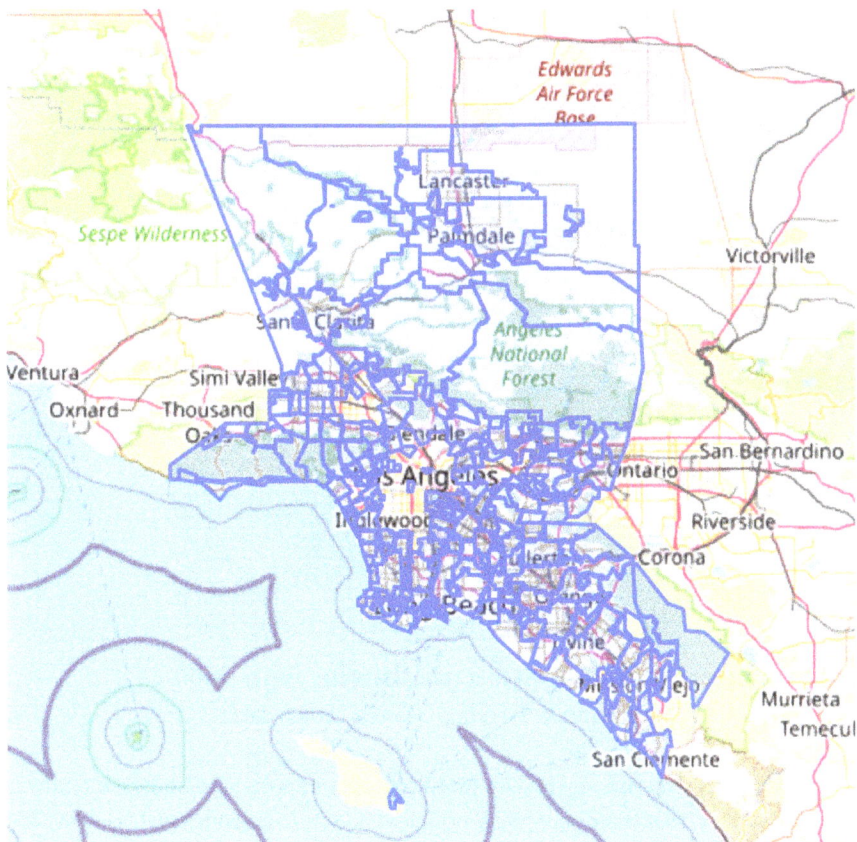

Figure 4.11 – Hispanic or Latino population percentage change from 2010
to 2020 (greater than the growth in the white population)

Better SQL queries begin with better questions. Group quarters data includes correctional facilities. You might want to know where these facilities are located and perhaps where there are more than 500 inmates. These examples are snippets of bigger questions but demonstrate how the displayed data reflect the filtering and specificity of your queries. *Figure 4.12* does not demonstrate clear patterns yet. What are a few additional datasets that you might consider including here?

```
SELECT geom,"name", ch4."gq_la".correctionals FROM ch4."gq_la"
WHERE ch4."gq_la".correctionals > 500;
```

Figure 4.12 – Location of correctional facilities with more than
500 inmates in the group quarters census data

When working with survey data and census data specifically, it is important to understand how the data is generated and what the variables are expressing in your analyses. User guides such as the 2020 Survey of Income and Program Participation (https://www2.census.gov/programs-surveys/sipp/tech-documentation/methodology/2020_SIPP_Users_Guide_OCT21.pdf) are important to understand the data definitions and how the data was weighted.

Briefly, weights are calculated to determine the number of people that each surveyed participant represents. Different members of a population may be sampled with a host of probabilities and response rates and this is adjusted by weights. For example, if the upper quartile weight is 7,000, this equates to a respondent representing 7,000 people in the sample.

# Running queries in QGIS

How would you look at Los Angeles County by actual percentages of Hispanic and Latino populations by tract? We don't have that information but we can add a column that will calculate the total for us.

Let's go back to the **Field calculator** tool in QGIS. The **Field calculator** tool is great for computing a new vector layer (leaving the underlying data untouched). The dataset lists populations as totals, so a simple equation was able to generate a new column showing the populations as percentages based on the underlying total population of the tract, as in *Figure 4.13*.

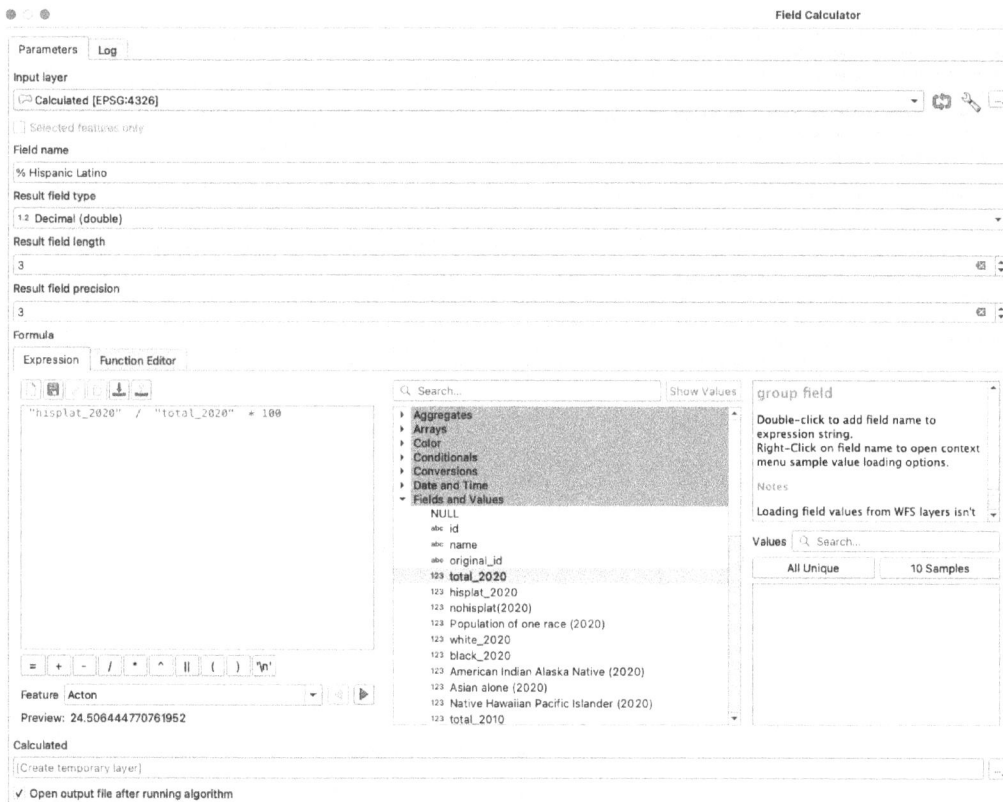

Figure 4.13 – The Field calculator tool creating a new column in QGIS

When you scroll down to **Fields and Values**, you will be able to select the fields and add the calculation in the **Expression** window. The new field will show up as a data layer in the canvas, labeled **Calculated**.

Run the query and you will be able to view your input parameters and verify the successful execution of the calculation:

```
- hisplat_2020/total_2020 * 100
QGIS version: 3.26.1-Buenos Aires
QGIS code revision: b609df9ed4
Qt version: 5.15.2
Python version: 3.9.5
GDAL version: 3.3.2
GEOS version: 3.9.1-CAPI-1.14.2
PROJ version: Rel. 8.1.1, September 1st, 2021
PDAL version: 2.3.0 (git-version: Release)
Algorithm started at: 2022-09-23T08:10:41
Algorithm 'Field calculator' starting…
Input parameters:
{ 'FIELD_LENGTH' : 3, 'FIELD_NAME' : '% Hispanic Latino',
'FIELD_PRECISION' : 3, 'FIELD_TYPE' : 0, 'FORMULA' : '
"hisplat_2020" / "total_2020" * 100', 'INPUT' : 'memory://Multi
Polygon?crs=EPSG:4326&field=id:string(-1,0)&field=name:string(-
1,0)&field=original_id:string(-1,0)&field=total_2020:integer(-
1,0)&field=hisplat_2020:integer(-1,0)&field=nohisplat%282020%29
:integer(-1,0)&field=Population%20of%20one%20race%20%28…
Execution completed in 0.16 seconds
Results:
{'OUTPUT': 'Calculated_0aea2a82_ba41_4c86_b7ad_8809977fd736'}
Loading resulting layers
Algorithm 'Field calculator' finished
```

We have already examined percentage increases in population totals and *Figure 4.14* now shows us the actual percentages in 2020 as reported by the census.

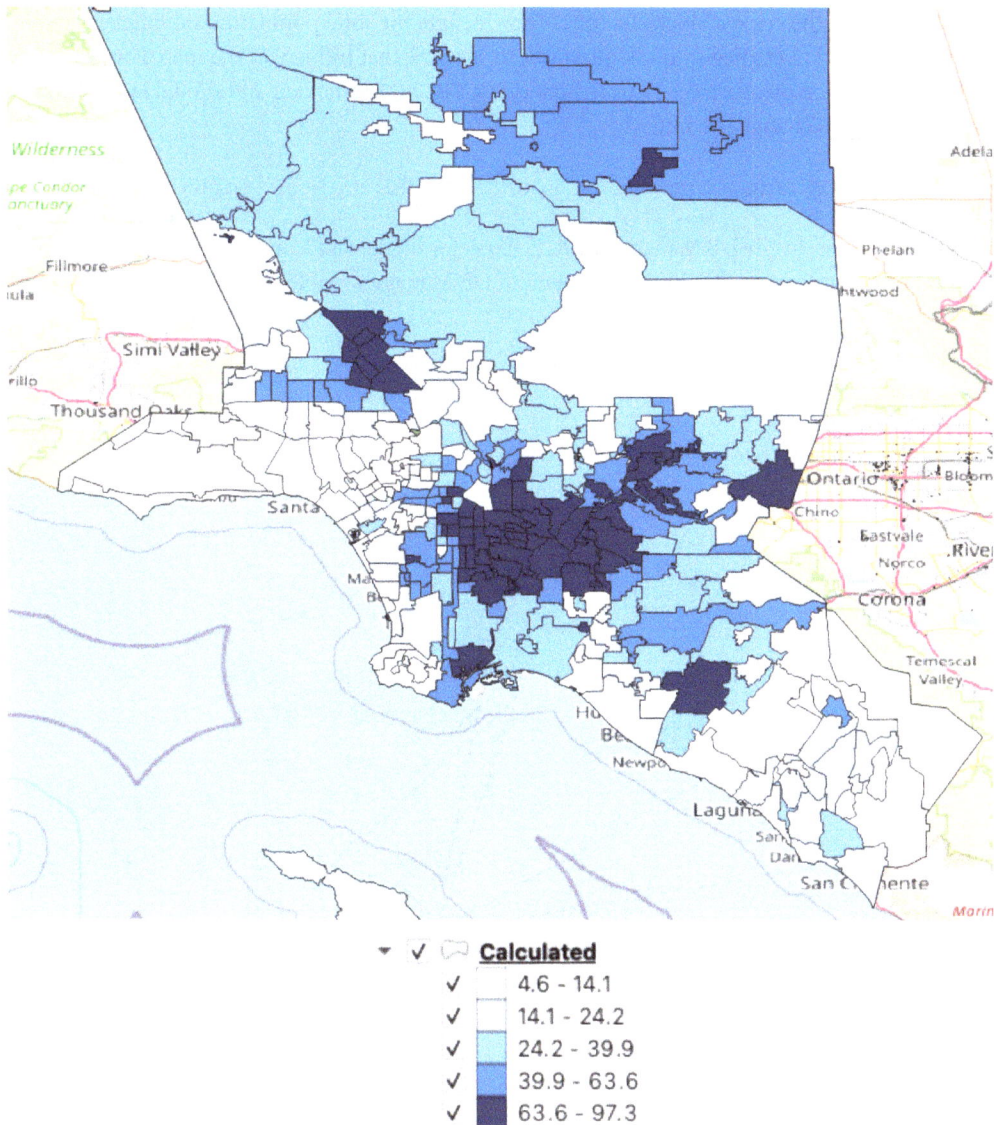

Figure 4.14 – Percentage of Hispanic and Latino populations in the 2020 census

Understanding that the work isn't done once the map is loaded onto the canvas is crucial to encouraging deeper inquiry and the application of spatial information. Descriptive statistics might ask where people live in Los Angeles County, California – but inferential statistics will be able to examine spatial relationships. We can make estimates on the probability that value x is more frequent in a location than value y. Working with census data, we can ask how variables are related in data over time in different locations. What might be influencing these findings?

*Figure 4.15* looks at the census blocks in 2020. They include the total population (and housing unit count) for each block. Let's begin to ask questions to see how that influences the map. For example, let's select a population greater than 200. The query is a SELECT statement and should look familiar. What happens if you change the values?

```
SELECT * FROM ch4."tiger20LA_block" WHERE "block_pop" > 200
```

You can run the query inside DB Manager in QGIS. Look for the second icon from the left in the figure to the right of the refresh symbol in the top menu of DB Manager, visible in *Figure 4.15*.

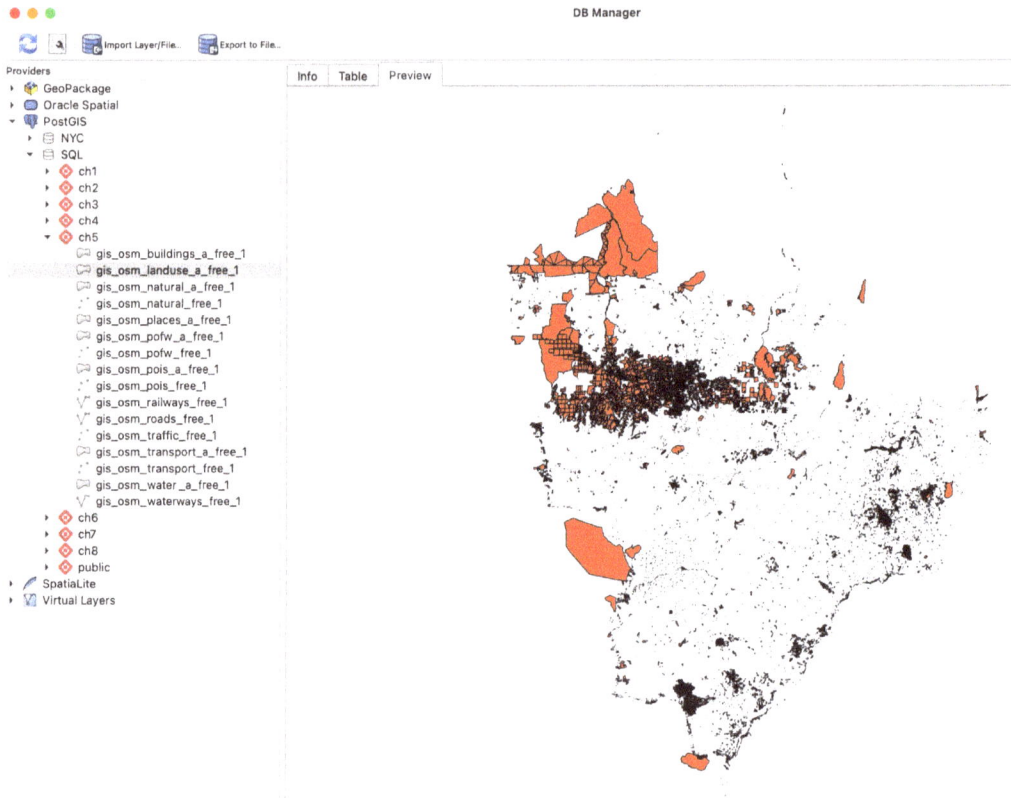

Figure 4.15 – Census blocks in Los Angeles County that contain more than 200 people

The census population data surveying how Hispanic and Latino populations have increased since the 2010 census can be viewed with a simple query. In QGIS, loading `hisplat_la` onto the canvas and viewing the **Layer Styling** Panel allows you to create intervals to capture blocks with a larger population of Hispanic and Latino communities as shown in *Figure 4.14*.

Running the following code in the SQL window in DB Manager, you can add the Query Layer and the `Fire_Hazard_Severity_Zones` data. Select the load as a layer option and click on **load**. Your new labeled layer (you can choose what to call it) is available on the **Layer** pane.

Are there patterns to observe where populations are located in relation to fire hazard severity? Using the **Layers** panel to classify the data by category, you can select the colors to display on the legend.

`Fire_Hazard_Severity_Zones` represent areas where mitigation strategies are in effect to reduce the risk of fires due to fuel, terrain, weather, and other causes. Adding data layers that allow you to compare risks across a geographical area and the nature of populations impacted by the risks and mitigation strategies can add an important dimensionality to measure equity and opportunity, as well as persistent barriers that might arise from the infrastructure built within and between communities.

First, select the layer you would like to style. Choosing the **Categorized** option and the column or value to highlight is the next step. The **Classify** button will provide you with options, as you now can view the number of categories in the dataset. In *Figure 4.16*, the symbol color was adjusted by highlighting the legend value and scrolling through the color profile. Any categories you do not want to highlight can be deleted by using the minus (-) symbol shown in red.

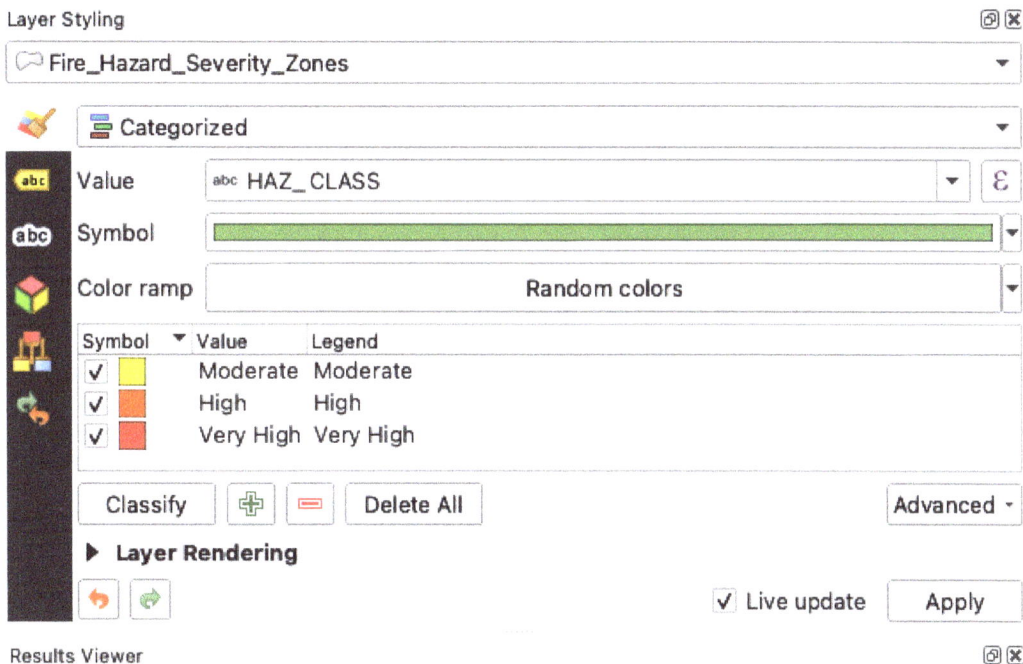

Figure 4.16 – Layer styling panel to create a color scheme for a Categorized feature class

*Figure 4.17* shows the visualization but the population level blends in with the **Moderate** version of `Fire_Hazard_Severity_Zones`. The ability to customize the features in the data layer is easily addressed in QGIS. When combining layers, you are able to adjust the color selection of symbols, and select different symbols (especially when working with point geometries).

```
SELECT hisplat_la.*


FROM ch4.hisplat_la,ch5."Fire_Hazard_Severity_Zones"
WHERE ST_Intersects (hisplat_la.geom,ch5."Fire_Hazard_Severity_
Zones".geom)
AND hisplat_la.hisplat_2020 > hisplat_la."nohisplat(2020)"
```

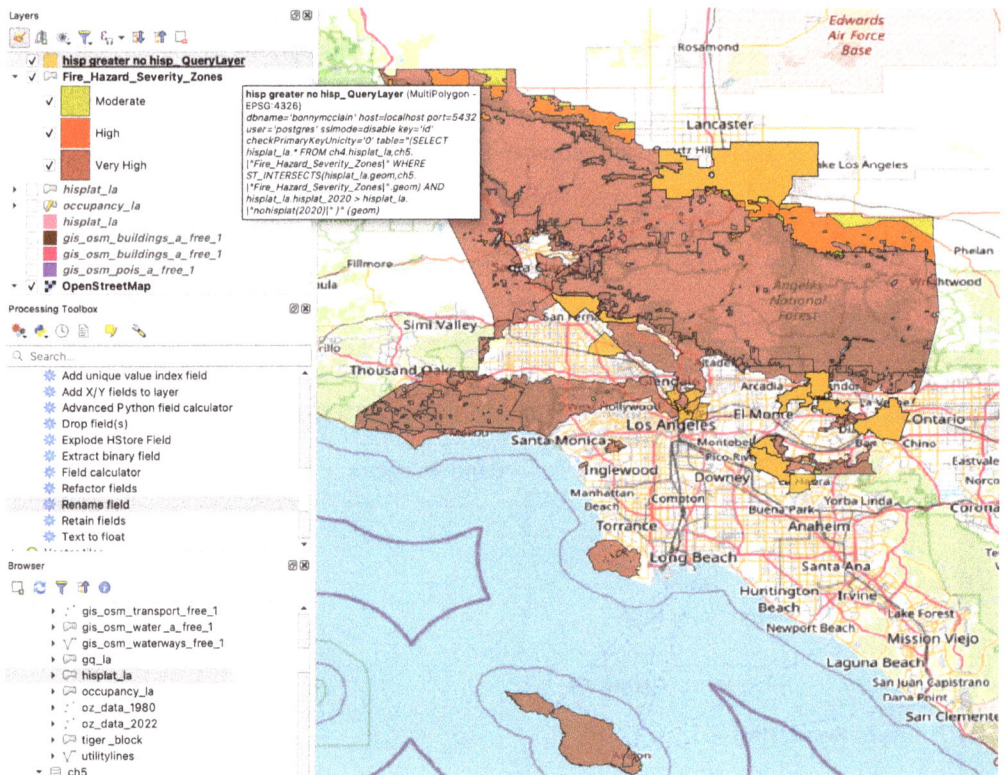

Figure 4.17 – Fire_Hazard_Severity_Zones and the location of populations
where Hispanic and Latino totals are greater than non-Hispanic

Explore the different options. More to come in later chapters.

The `ST_Intersects` function is indicating where Hispanic and Latino Populations (increased totals since the 2010 census) are located along fire zones. As shown in *Figure 4.18*, selecting a contrasting color is more impactful. Intersections between the polygons are queried and visualized in QGIS. Although they seem similar, several different methods provide quite different results in how the data is queried.

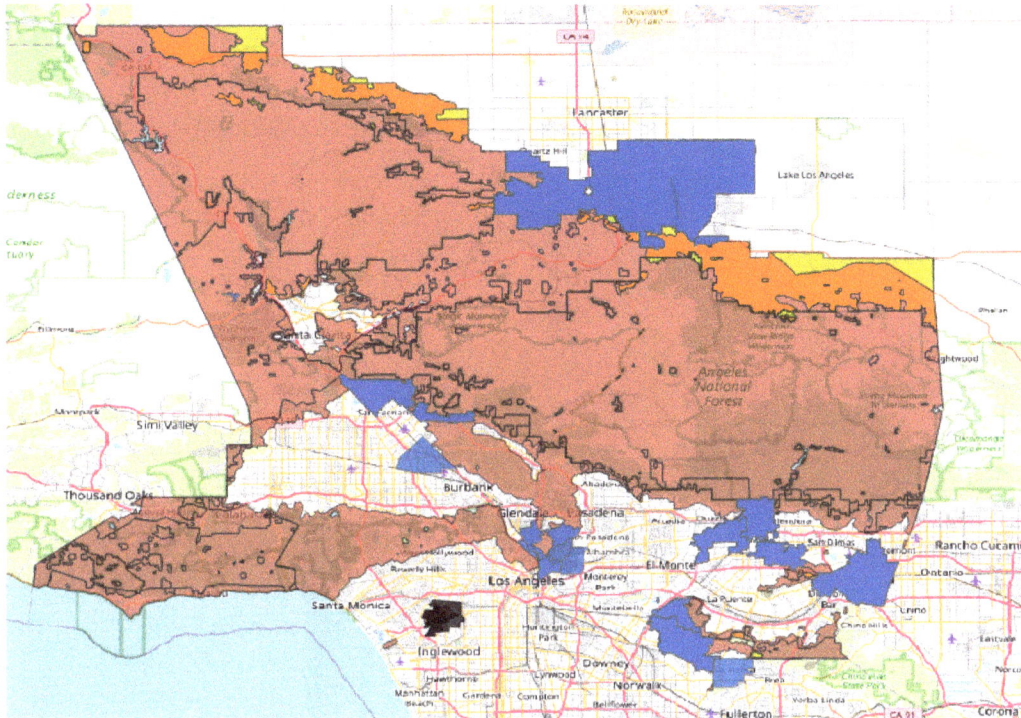

Figure 4.18 – Where increasing populations of Hispanic and Latino
people are living in relation to the fire hazard risk

Explore a few of the different topological relationship options, such as the following:

- `ST_Touches` (interiors of the polygons do not overlap)
- `ST_Contains` (the entire polygon is inside the other polygon)
- `ST_Crosses` (some points in common)

How does this impact the selected data?

Another spatial method is `ST_Centroid`. The geometries of the `geom` column in your table are collected and the geometric center of the mass is computed. When the geometry is a point or collection of points, the arithmetic mean is calculated. LineString centroids use the weighted length of each segment, and polygons compute centroids in terms of area.

*Figure 4.19* has a red star indicating the centroid. We will explore additional measures of centrality in the next chapter. They are also used when calculating distance and zone attributes, and often represent the polygon as a whole in advanced calculations:

```
SELECT ST_AsText(ST_Centroid(ST_Collect(geom))) FROM ch4.gis_
osm_landuse_a_free_1
Output:"POINT(-117.21356029457164 34.528429645559925)"
```

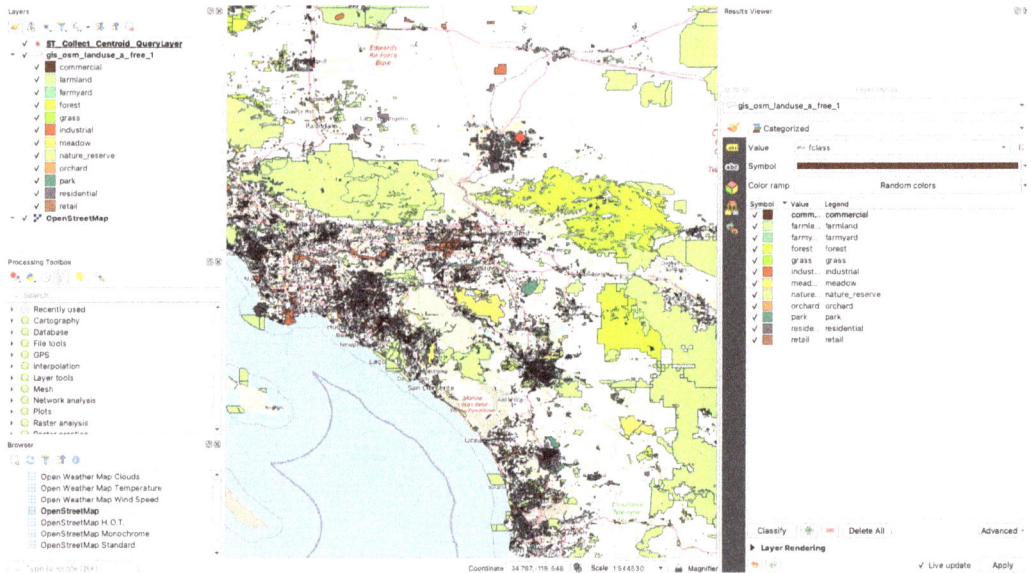

Figure 4.19 – Calculating the centroid in the OSM land use table

Typical spatial statistical concepts are easiest to understand when accessed to answer spatial queries. You will notice patterns and clusters of location data. Can we predict the unknown from the known information? Descriptive statistics can tell us where the fire zones are located in California. Do fires occur more frequently in certain locations? As we begin to explore more advanced SQL queries, it is important to determine whether we are analyzing one feature alone or whether the location is influenced by other attribute values.

So far, we have been working with vector data, mostly discrete in nature, but spatially continuous data (raster) is also common. In this chapter, you also worked with spatially continuous categorical data. The land cover types are assigned to categories as observed in the legends of the generated maps.

Census data is summarized as continuous data aggregated according to arbitrarily assigned geographic locations such as census tracts or blocks. Noticing that there are a lot of categorically similar populations in a census tract invites the exploration of discovering how they are related over time and how this differs across different tracts.

A good practice, to begin with, is to assume the null hypothesis – no pattern exists. Applying statistical principles, we should then have a high level of confidence that what we are observing is not due to chance. This is important because thematic maps can be manipulated. Examining the center, dispersion, and directional trends in our data is an important element of spatial analysis.

In later chapters, the mean center (geographic distributions), the median center (relationships between features), and the central feature will also measure the relationships between geographic distributions, measures of central tendency, and localized events. We can also measure clusters of, say, census tracts based on attribute values using hot spot analysis. While heat maps measure frequencies of events, hot spot analysis looks for statistical fixed distance bands.

## Building prediction models

When beginning to analyze data spatially, there are a few practices that will make the endeavor run more smoothly. It is always important when bringing together more than one dataset to evaluate the geometry columns. Running the following code will give you a glimpse into the SRID and data type in *Figure 4.20*:

```
SELECT Find_SRID('ch4','hisplat_la','geom');
SELECT * FROM geometry_columns
```

| f_table_catalog character varying (256) | f_table_schema name | f_table_name name | f_geometry_column name | coord_dimension integer | srid integer | type character varying (30) |
|---|---|---|---|---|---|---|
| bonnymcclain | ch3 | attains_au_lin... | geom | 2 | 3857 | MULTILINESTRING |
| bonnymcclain | ch2 | buildingFP | geom | 2 | 4326 | MULTIPOLYGON |
| bonnymcclain | ch2 | streetNYC | geom | 2 | 4326 | MULTILINESTRING |
| bonnymcclain | ch2 | streetrating | geom | 2 | 4326 | MULTILINESTRING |
| bonnymcclain | ch2 | buildingp | geom | 2 | 4326 | POINT |
| bonnymcclain | ch3 | dohmh | geom | 2 | 4326 | POINT |
| bonnymcclain | ch4 | occupancy_la | geom | 2 | 4326 | MULTIPOLYGON |
| bonnymcclain | ch4 | gis_osm_tran... | geom | 2 | 4326 | POINT |
| bonnymcclain | ch4 | utilitylines | geom | 2 | 4326 | LINESTRING |
| bonnymcclain | ch4 | gis_osm_nat... | geom | 2 | 4326 | MULTIPOLYGON |
| bonnymcclain | ch4 | gis_osm_plac... | geom | 2 | 4326 | MULTIPOLYGON |
| bonnymcclain | ch3 | attains_au_ca... | geom | 2 | 3857 | MULTIPOLYGON |
| bonnymcclain | ch5 | gis_osm_buil... | geom | 2 | 4326 | MULTIPOLYGON |
| bonnymcclain | ch6 | blackgreater_... | geom | 2 | 4326 | MULTIPOLYGON |

Figure 4.20 – Reviewing the table catalog in pgAdmin

Understanding how to query datasets with SQL is the first step in introducing prediction models to the database. The execution of these properties will be expanded on as our journey continues.

Exploring the `below_poverty_censustract` data for Los Angeles County, I want to be able to isolate a tract and explore neighboring tracts. Location and distance might hold clues for exploring marginalized communities or populations living below the poverty line.

Looking for values to explore, I am exploring tracts with a high percentage of the population living below the poverty level. Running the following code to order the data by descending values will help me to find the higher percentages:

```
SELECT * FROM ch4.below_poverty_censustract
ORDER BY below_fpl_pct DESC
```

I identified the tract of interest by tract number and queried any tracts that intersect. `ST_Intersects` compares two geometries and returns `true` if they have any point in common. The following code was written into the query editor in QGIS. Execute the code and name the query layer in the box in *Figure 4.21*:

```
SELECT * FROM ch4.below_poverty_censustract
WHERE ST_Intersects (ch4.below_poverty_censustract.geom,
(SELECT geom
FROM ch4.below_poverty_censustract
WHERE "tract" = '06037980014'))
```

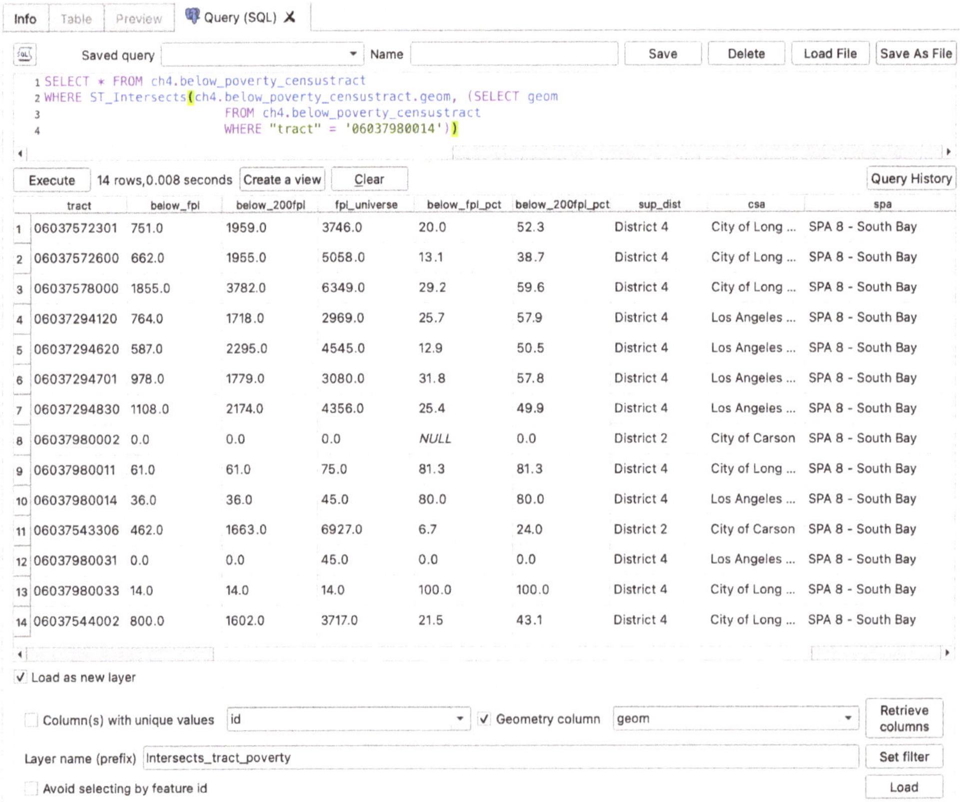


| | tract | below_fpl | below_200fpl | fpl_universe | below_fpl_pct | below_200fpl_pct | sup_dist | csa | spa |
|---|---|---|---|---|---|---|---|---|---|
| 1 | 06037572301 | 751.0 | 1959.0 | 3746.0 | 20.0 | 52.3 | District 4 | City of Long ... | SPA 8 - South Bay |
| 2 | 06037572600 | 662.0 | 1955.0 | 5058.0 | 13.1 | 38.7 | District 4 | City of Long ... | SPA 8 - South Bay |
| 3 | 060375780000 | 1855.0 | 3782.0 | 6349.0 | 29.2 | 59.6 | District 4 | City of Long ... | SPA 8 - South Bay |
| 4 | 06037294120 | 764.0 | 1718.0 | 2969.0 | 25.7 | 57.9 | District 4 | Los Angeles ... | SPA 8 - South Bay |
| 5 | 06037294620 | 587.0 | 2295.0 | 4545.0 | 12.9 | 50.5 | District 4 | Los Angeles ... | SPA 8 - South Bay |
| 6 | 06037294701 | 978.0 | 1779.0 | 3080.0 | 31.8 | 57.8 | District 4 | Los Angeles ... | SPA 8 - South Bay |
| 7 | 06037294830 | 1108.0 | 2174.0 | 4356.0 | 25.4 | 49.9 | District 4 | Los Angeles ... | SPA 8 - South Bay |
| 8 | 06037980002 | 0.0 | 0.0 | 0.0 | NULL | 0.0 | District 2 | City of Carson | SPA 8 - South Bay |
| 9 | 06037980011 | 61.0 | 61.0 | 75.0 | 81.3 | 81.3 | District 4 | City of Long ... | SPA 8 - South Bay |
| 10 | 06037980014 | 36.0 | 36.0 | 45.0 | 80.0 | 80.0 | District 4 | Los Angeles ... | SPA 8 - South Bay |
| 11 | 06037543306 | 462.0 | 1663.0 | 6927.0 | 6.7 | 24.0 | District 2 | City of Carson | SPA 8 - South Bay |
| 12 | 06037980031 | 0.0 | 0.0 | 45.0 | 0.0 | 0.0 | District 4 | Los Angeles ... | SPA 8 - South Bay |
| 13 | 06037980033 | 14.0 | 14.0 | 14.0 | 100.0 | 100.0 | District 4 | City of Long ... | SPA 8 - South Bay |
| 14 | 06037544002 | 800.0 | 1602.0 | 3717.0 | 21.5 | 43.1 | District 4 | City of Long ... | SPA 8 - South Bay |

Figure 4.21 – SQL query in QGIS DB manager

Remember the geometry viewer of the single tract visible in pgAdmin before opening QGIS. The nature of the neighborhood is visible and we can see that it is near the Long Beach Terminal. It is the blue polygon in *Figure 4.22*.

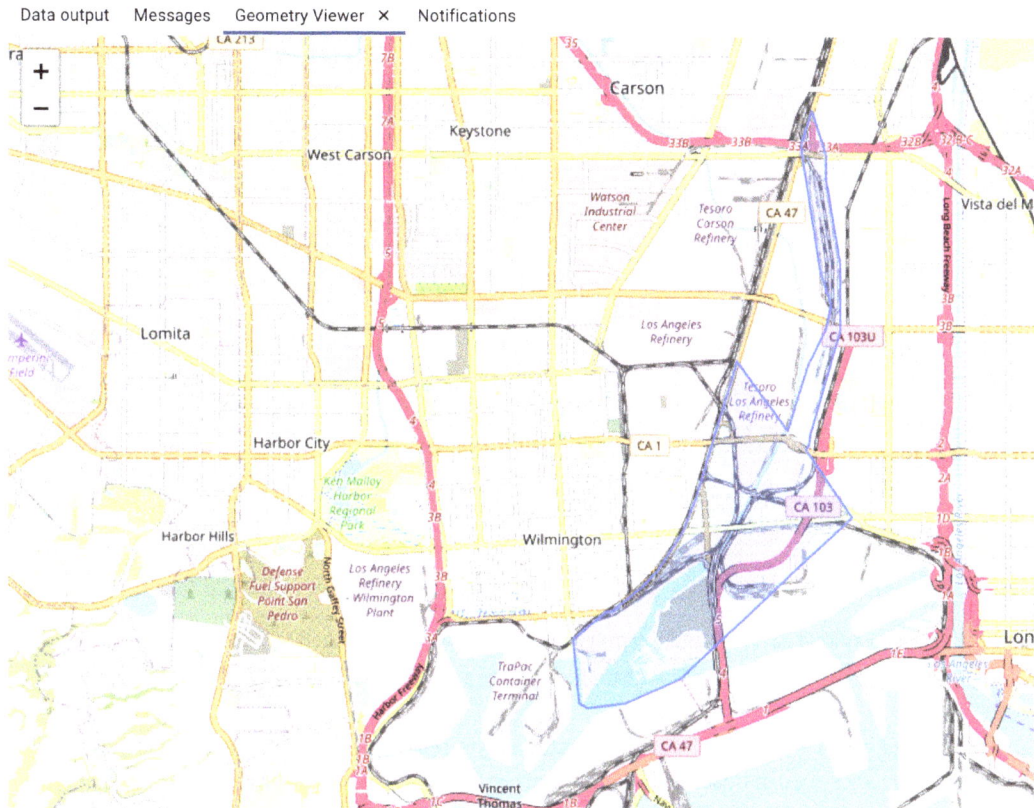

Figure 4.22 – Geometry viewer displays the selected tract in pgAdmin

How can we identify areas adjacent to or near our tract of interest? Learning about a few of the spatial methods will be helpful in later chapters where we will be able to predict output values by examining a wide variety of features of the locations surrounding our outcomes of interest. Visible in *Figure 4.23*, we can see proximity to water, income level, and other features that we will examine using regression experiments or time-series predictions in later chapters. For now, begin asking questions. What type of data would you like to see?

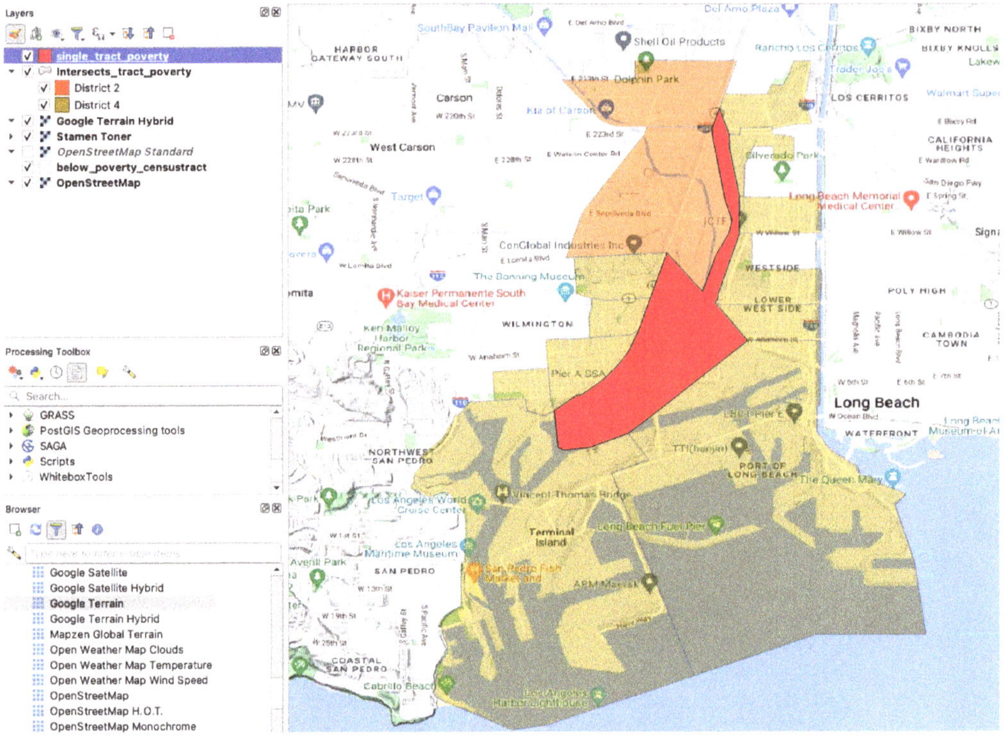

Figure 4.23 – Intersecting polygons organized by Los Angeles County
Board of Supervisor districts and high percentages of poverty

Many of the data layers we can explore are included in the census we began looking at earlier in the chapter. In *Figure 4.23*, we can observe different tracts and explore how their poverty rates compare to the initial tract we selected. Different county supervisors can be identified categorically in the **Layers** panel, and revisiting the census tables created in QGIS and imported to pgAdmin would likely be a useful next step.

Areas such as Los Angeles County are prone to fires, landslides, and a host of other risks of natural disasters, *figure 4.24*.

Figure 4.24 – Landslide zones in Los Angeles County

The complexity of looking at the topography and how it influences the demographics of people and infrastructure, especially in an area such as Los Angeles, California, becomes an opportunity to engage with geospatial tools and ask bigger questions.

*Figure 4.25* is a zoomed-in map of the landslide zones in and surrounding the urban area of Los Angeles. Although we will be leaving the California area in the next chapter, the objective here has been to introduce you to a few data portals where publicly available datasets are available for exploration. The data is from the county of Los Angeles – `https://data.lacounty.gov/datasets/ lacounty::landslide-zones/about` – and depicts areas prone to landslides. If you are like me, perhaps coastal communities mostly came to mind, and you didn't imagine the risk existing in urban areas as well.

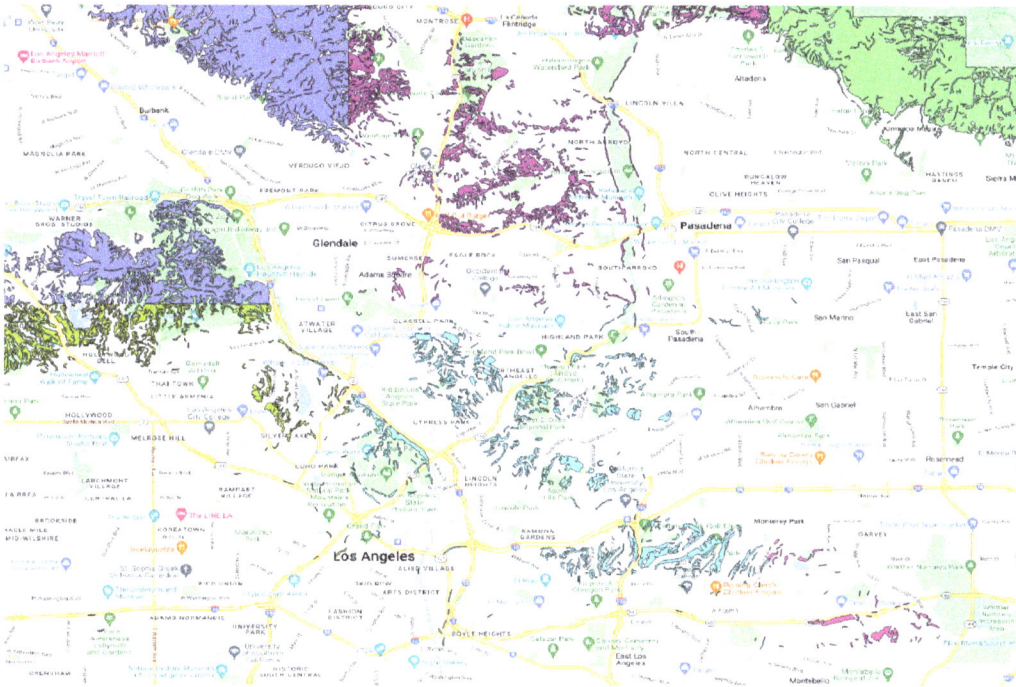

Figure 4.25 – Landslide zones near Los Angeles, California

Let's revisit the fire zones as we close out the chapter. Before we can build prediction models, it is important to gain a sense of the type of questions we might formulate.

In *Figure 4.26* here, the building data from **OSM** is added as a layer to the canvas. It isn't fully loaded, as the density of buildings would obscure the detail I am trying to highlight at this zoom setting.

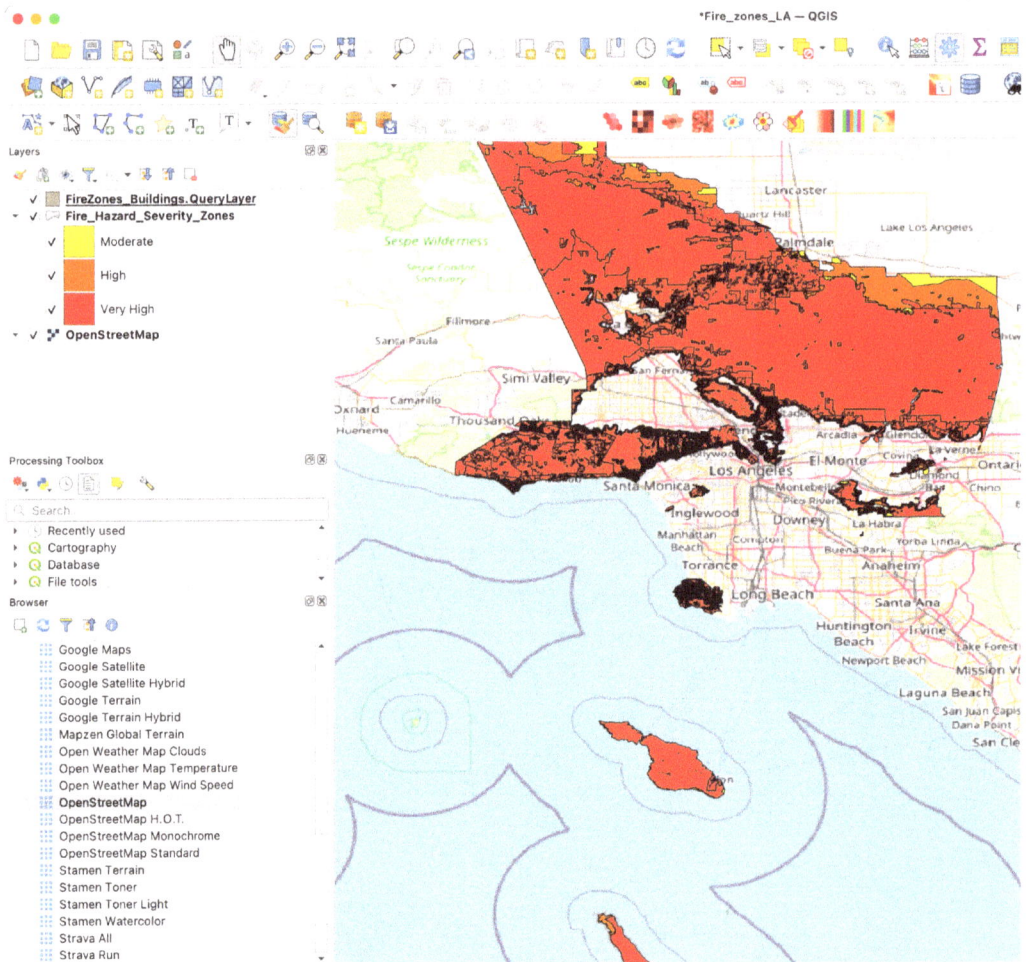

Figure 4.26 – Buildings within fire hazard severity zones in QGIS

My question to you is do you think there is a pattern to where certain buildings are located in these higher-risk zones? Can we examine demographic information from the census and hypothesize and test these hypotheses?

I hope you have an appreciation of the powerful integration capabilities of working with SQL in both the pgAdmin interface and with QGIS. *Figure 4.26* highlights the utility of the visualization interface, styling capabilities, and the ability to run SQL queries right in the canvas. By generating query layers, you are able to visualize the results in real time. Become curious about your basemap selection, the transparency of different layers, and the other customizable features.

Let's summarize the chapter.

# Summary

In this chapter, you learned how to write a few spatial topographic queries and add multiple layers of data to your SQL queries. You were introduced to an important resource for understanding demographic and economic data attributes associated with geographical locations – the 2020 US Census. You successfully ran queries on data and generated your own maps for exploration.

In *Chapter 5*, *Using SQL Functions – Spatial and Non-Spatial*, you will continue to be introduced to SQL spatial queries and begin to work with more advanced concepts, such as measuring distance and evaluating buffers, in addition to building a simple prediction model.

# Section 2:
# SQL for Spatial Analytics

In *Part 2*, readers will learn about spatial data types and how to abstract and encapsulate spatial structures such as boundaries and dimensions. Finally, large datasets are then enabled for local analyses. Open source GIS, combined with plugins that expand QGIS functionality, have made QGIS an important tool for analyzing spatial information.

This section has the following chapters:

- *Chapter 5, Using SQL Functions – Spatial and Non-Spatial*
- *Chapter 6, Building SQL Queries Visually in a Graphical Query Builder*
- *Chapter 7, Exploring PostGIS for Geographic Analysis*
- *Chapter 8, Integrating SQL with QGIS*

# 5
# Using SQL Functions – Spatial and Non-Spatial

Beginning with a data question is the most efficient process for writing SQL queries. This may be a problem you are trying to solve or perhaps a null hypothesis. Once we have an articulated question, we will be tasked with locating data. The final steps are to analyze the problem and communicate the results.

I would argue that this is particularly relevant when learning how to write efficient SQL queries. There is a natural cadence and syntax that differs from writing a code snippet in, say, Python or R.

The datasets we are exploring are from the Amazon rainforest in Brazil. Mining activities in the Amazon rainforest are a known provocation for deforestation and the associated impact of toxic pollution on surrounding communities. Mining activities also require dense road networks for transportation and infrastructure to support industrial mining, which impacts the growth of surrounding vegetation, pollution run-off, and habitats. The areas of Brazil with the densest vegetation and risk of deforestation are often inhabited by indigenous populations, which are often the people with the least influence to restrict access and defend protected lands.

The following topics will be covered in this chapter:

- Exploring spatial statistics to explore distance, buffers, and proximity to characteristics of the land, populations, and different attributes. These are represented by points, multi-lines, and polygons.

- An introduction to writing user-defined functions.

By the end of this chapter, you will understand how to write SQL queries and explore both spatial and non-spatial data.

## Technical requirements

I invite you to find your own data if you are comfortable or access the data recommended in this book's GitHub repository at https://github.com/PacktPublishing/Geospatial-Analysis-with-SQL.

We will access the following data to explore the questions in this chapter:

- Buildings in Brazil (places) and **points of interest** (**POIs**)
- Indigenous population boundary polygons
- Mining sites in Brazil
- Roads

# Preparing data

So far, you have learned how to download data directly from Open Street Map and begin asking data questions. When you prefer to work with a smaller dataset, instead of downloading larger files directly to your computer, you can also create datasets directly from the map canvas and write SQL queries entirely in QGIS.

QGIS has a training dataset but it is not too complicated to substitute the default set with something of more interest. We can select a region of interest from the canvas and explore different layers by using a QGIS plugin known as **QuickOSM**.

There is also a plugin for uploading data into the QGIS canvas for exploring our region of interest.

## Creating a dataset using the Open Street Map plugin

Here is how to install the QuickOSM plugin:

1.  To install the QuickOSM plugin, navigate to the menu at the top of the canvas and search for QuickOSM in the pane, as shown in *Figure 5.1*. As you can see, I already have the plugin installed; you will notice a button visible so that you can install it.

2.  Next, open a new project in QGIS and navigate to **DB Manager**. As a reminder, you can access it by selecting **Layer** from the top menu bar. Scroll down in the **Browser** window and select **OpenStreetMap**. Alternatively, you can open the base map directly in the **Browser** window:

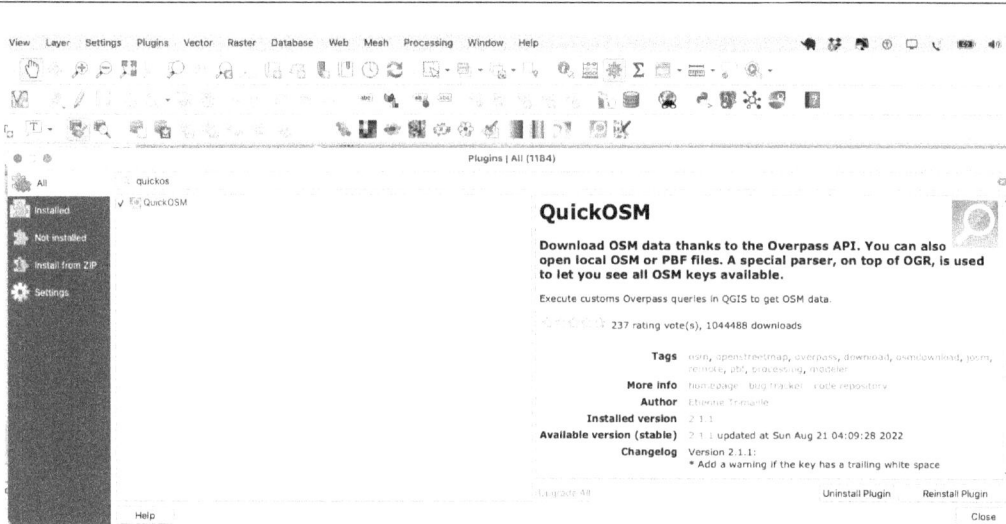

Figure 5.1 – QuickOSM plugin in QGIS

3. Return to the canvas and go to the **OpenStreetMap** base map

4. The lower left corner of the QGIS console has a search function shown in *figure 5.2*. Upon clicking inside the search box, you will see **Nominatim Geocoder**, where you can enter locations of interest directly instead of searching for and zooming in on locations. Select the Nominatim Geocoder and enter the location in the search box. If you type `Palmas`, the map will provide options and then take you to the location you entered.

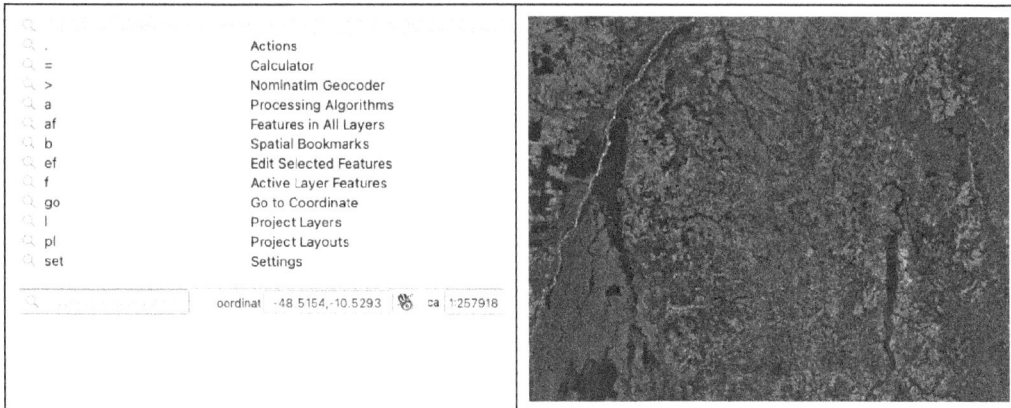

Figure 5.2 – OpenStreetMap base map

We are exploring Brazil and the Amazon rainforest. There are many areas of interest. For this example, we are zooming in on **Palmas** in the **STATE OF TOCANTINS**. It is in the southeast portion of Brazil that's visible in *Figure 5.2* as a small red dot. I selected this area as there is a lot of activity, but feel free to look around. You will most likely need to zoom in as the plugin will timeout over larger regions (you can reset this from the default) and it will be easier to simplify the introduction to using this tool.

You should notice the little icon (a magnifying glass on a green square) in your toolbar, as visible in *Figure 5.1*:

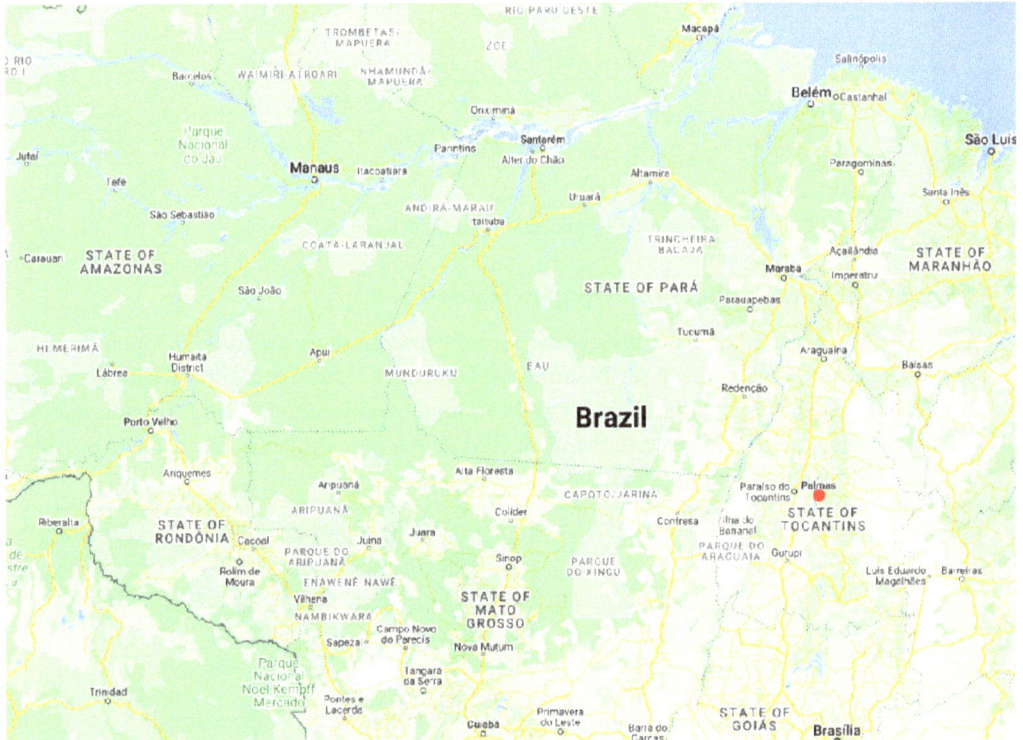

Figure 5.3 – Locating an area of interest in Open Street Map in QGIS

5.   When you execute the **QuickOSM** tool from the toolbar or the **Vector** pill in the menu, you will see a **Quick** query tab with a bunch of presets that you can select. Scroll through the options in *Figure 5.4* by clicking the **Preset** dropdown.

Because we are specifically looking for roads and buildings, these are the queries I selected in *Figure 5.4* and *Figure 5.5*. Select **Canvas Extent** so that the data is only downloaded from your specific area of interest. Run the query; the data layer will be visible in the canvas. Be sure to also indicate the geometry you are exploring.

6.  Go to the **Advanced** section and check the appropriate boxes on the right. **Multipolygons** should also be selected. Update these settings when your data is points or lines. You can also reset the **Timeout** option to accommodate larger files. In *figure 5.4* it defaults to 25 but I routinely set it to 2000. Run the query:

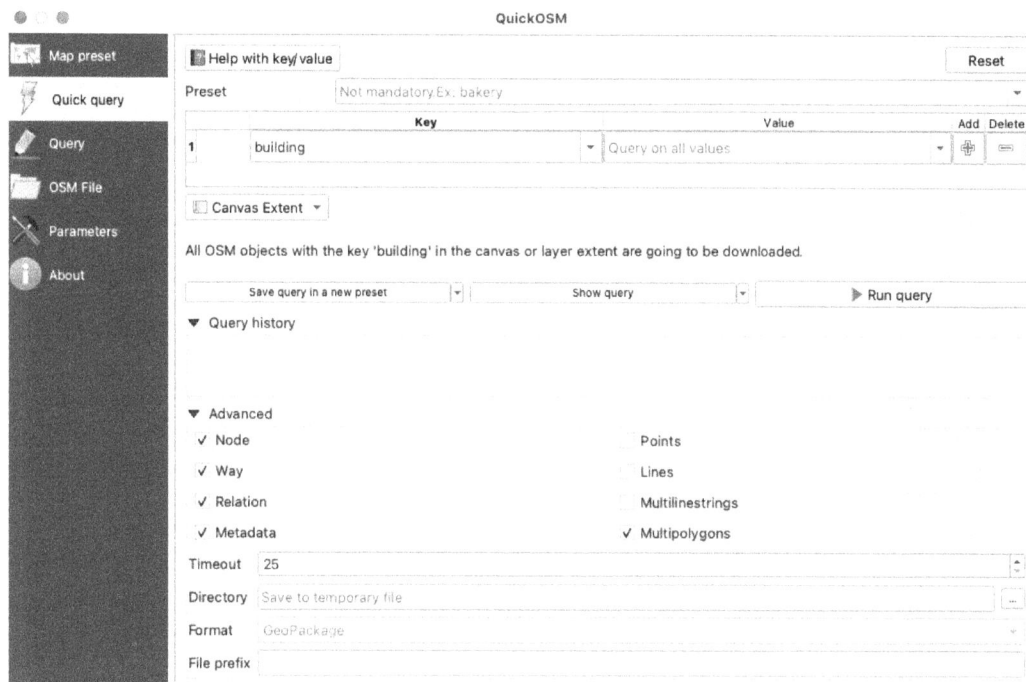

Figure 5.4 – Creating the building data layers in QuickOSM

7.  Next, we will create an additional data layer for roads, as shown in *Figure 5.5*, but remember to select **Multilinestrings** and **Lines** instead of **Multipolygons**. Repeat this process for layers in individual projects that interest you! Here is a resource for understanding OSM features: https://wiki.openstreetmap.org/wiki/Map_features. Expand the **Advanced** section to see the rest of the available options:

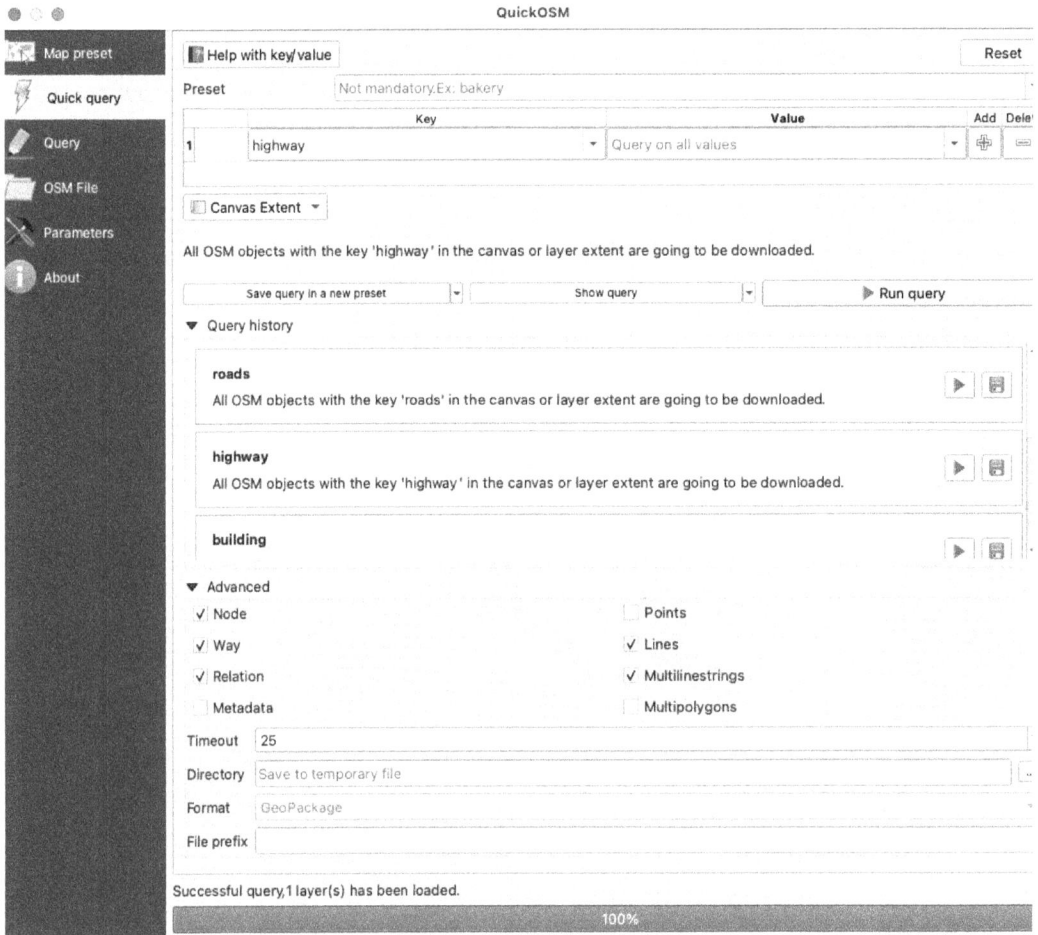

Figure 5.5 – QuickOSM – selecting a highway key in QGIS

8.  The layers you add using this method are temporary. The little rectangular or indicator icon visible next to the **place** layer in *Figure 5.6* indicates a scratch layer:

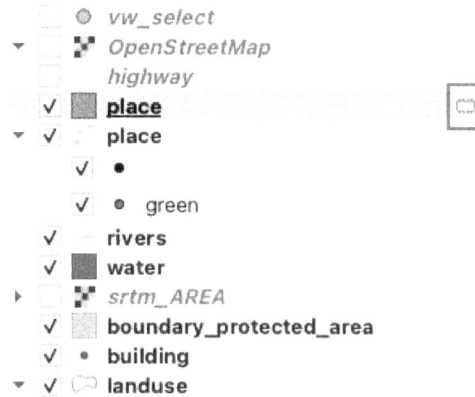

Figure 5.6 – Scratch layer with an indicator icon

9.  Clicking on this icon will bring you to the window shown in *Figure 5.7*. There are many options for saving the layer. I lean toward **GeoJSON**, but **ESRI Shapefile** is popular as well. **GeoPackage** is a standards-based data format (https://www.geopackage.org):

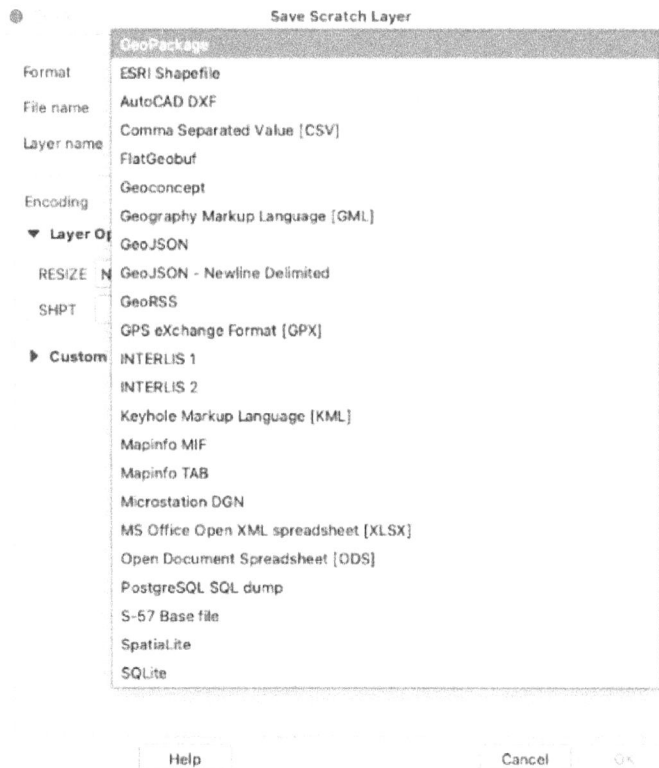

Figure 5.7 – Saving a scratch layer in QGIS

10. Upon saving a scratch layer, you will need to enter a few details when prompted, including **Layer name** and other options; refer to *Figure 5.8*. Add different layers until you have a decent variety. You can use the QGIS sample data (on GitHub) or build your own while following the vector file types as a guide:

Figure 5.8 – Saving a scratch layer in QGIS

To save the data, we have created a folder called `practice_data` in `Downloads` and added all the `Layer` instances into that folder, being careful to save them in the correct formats.

This was a brief introduction to how you can customize your data. I think it is important to have a variety of ways to achieve something such as finding and bringing your data into QGIS or your SQL database. This is the data we will use for the rest of this chapter.

But what do you do if data is missing from the dataset?

## Null values in PostgreSQL

Before we dive into analysis, it is important to understand what null signifies and what we can do with it. Briefly, in Postgres, NULL has no value. It doesn't equal 0, so you can't use it in mathematical calculations. In a nutshell, it indicates a field without a value. In large datasets, it makes sense that not all columns contain data. In the case of OSM, perhaps that street has not been labeled with a name or the identity of that building is unknown. It doesn't mean the value is nothing but that it is unknown. In the PostgreSQL database, there are different scenarios where NULL is the expected value – this differs in each coding language, so be vigilant (and curious).

The data we will use for the rest of this chapter can be found in the GitHub repository mentioned earlier and includes `Rondinia.gdb`. As a reminder, we will locate the downloaded files in the **Browser** area, add them to the canvas, and bring them into PostgreSQL using **DataBase Manager**, as seen in *Figure 5.9*. We have used numbered `schema` for tables here but you may also simply use the `Public` schema:

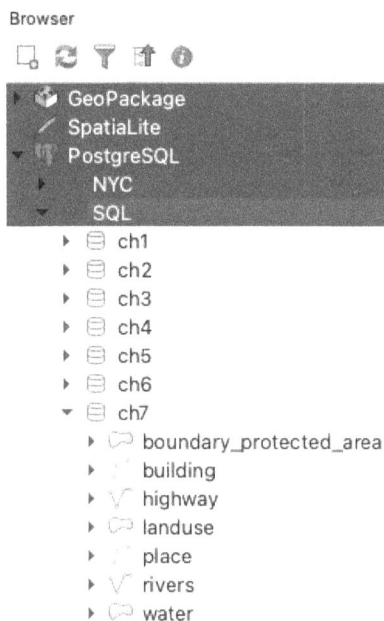

Figure 5.9 – Locating data files in PostgreSQL

Remember to refresh your files to update the schemas and tables.

# Spatial concepts

Understanding how spatial statements are constructed and how to customize them for your queries is an important task in learning new ways to interact with spatial datasets.

## ST_Within

The synopsis of `ST_Within` includes the following:

```
boolean ST_Within(geometry A, geometry B);
```

A true and false determination is provided to determine if `geometry A` is entirely within `geometry B`.

You will need to be comfortable with SELECT statements. Consider the following code. I am interested in buildings located within protected areas. I need to SELECT the buildings and would like to identify them by their name and fid. Geopackages often use fid as their unique identifier, although it isn't often necessary since OSM has its own identification for its features. I grabbed it here as an example. Any unique identifier will work. I ran the following query in pgAdmin.

The name, fid, and geometry (point) are displayed in the output window once you run the code. ST_AsText is returning the **Well-Known Text (WKT)** format of the geometry data to standardize the format returned:

```
SELECT ch5.building.name, ch5.building.fid, ST_AsText(ch5.
building.geom) as point
FROM ch5.building,ch5.boundary_protected_area
WHERE ST_Within(ch5.building.geom, ch5.boundary_protected_area.
geom)
```

The tables I am accessing are listed in the FROM statement. Next, I added the condition to include only the rows where the buildings are located within the protected areas of *Figure 5.9*. The geometry of the building is only returned if it is entirely within the geometry of the protected area. Anything spanning the geometry will not be included.

## CREATE VIEW

We can also CREATE VIEW. I often use this when writing queries in pgAdmin but plan on visualizing them in QGIS. This is the view of a specific query. You can locate the **Layer** property in the schema where it was created or in your Public schema folder. The format requires you to indicate a **VIEW** name and your query (AS is a reserved SQL keyword) follows this example. The query indicates which columns to include. Run the following code:

```
CREATE VIEW vw_select_location AS
   SELECT ch5.building.fid, ch5.building.name, ch5.building.geom
     FROM ch5.building, ch5.boundary_protected_area
       WHERE ST_Within(ch5.building.geom, ch5.boundary_
protected_area.geom)
```

The output is shown in *Figure 5.10*. Select the downward arrow and download the output to your computer. The output is formatted as a CSV and you can upload it into QGIS by using the **Data Source Manager** area or selecting **Add Layer** from the **Layer** menu. Simply drag the layer onto your canvas and it will be added to the map:

Data output    Messages    Notifications

| name character varying | fid bigint | point text |
|---|---|---|
| 42  [null] | 406 | POINT(-47.7792054 -15.9016729) |
| 43  [null] | 407 | POINT(-47.7788285 -15.9021683) |
| 44  Portaria do Solar ... | 37 | POINT(-47.7644607 -15.8501053) |
| 45  Centro Empresari... | 102 | POINT(-47.8222678 -15.8720196) |
| 46  Villa Patrícia Eve... | 123 | POINT(-47.8122491 -15.8700304) |
| 47  Congregação Cri... | 124 | POINT(-47.7078562 -15.6857003) |
| 48  [null] | 284 | POINT(-47.8212041 -15.86854) |
| 49  [null] | 287 | POINT(-47.8213366 -15.8691448) |
| 50  [null] | 288 | POINT(-47.8214666 -15.8695569) |
| 51  Centro Hípico La... | 339 | POINT(-47.8107205 -15.8520667) |
| 52  [null] | 364 | POINT(-47.8151844 -15.8638283) |
| 53  [null] | 370 | POINT(-47.7647463 -15.7359266) |
| 54  [null] | 371 | POINT(-47.766792 -15.7372579) |
| 55  [null] | 385 | POINT(-47.7779918 -15.6529) |

Figure 5.10 – CREATE VIEW displaying buildings located within protected areas

The **Geometry Viewer** area in pgAdmin is convenient for looking at maps but only if you change the SRID value. In *Figure 5.11*, I have selected **Geometry Viewer** just to locate the data, but let's head over to QGIS for a better view without having to write more code to change the projection:

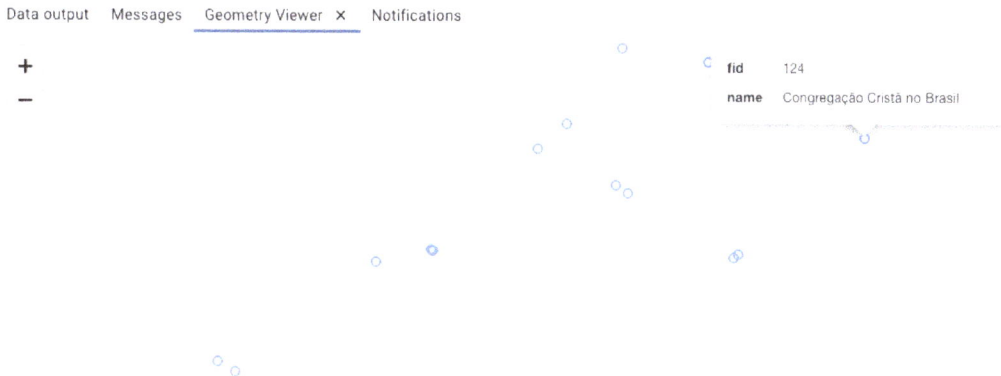

Figure 5.11 – Visualizing CREATE VIEW in Geometry Viewer

Scroll down the **Browser** area in QGIS and select the layers to add to your view for this example. Running SQL queries in QGIS also allows you to generate a **View Layer**. Alternatively, you can simply run the code and select **Create a view** from the menu, as displayed in *Figure 5.12*:

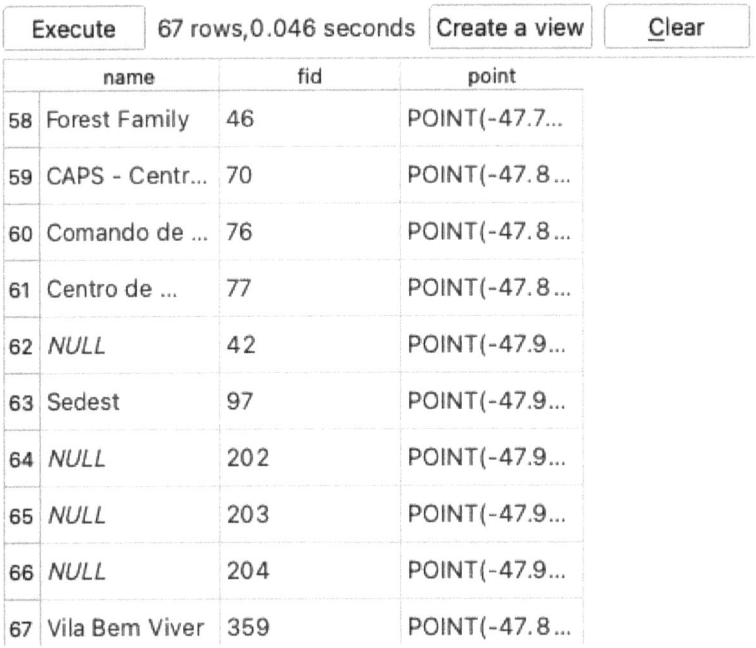

| Execute | 67 rows, 0.046 seconds | Create a view | | Clear |
| --- | --- | --- | --- | --- |
| | name | fid | point | |
| 58 | Forest Family | 46 | POINT(-47.7... | |
| 59 | CAPS - Centr... | 70 | POINT(-47.8... | |
| 60 | Comando de ... | 76 | POINT(-47.8... | |
| 61 | Centro de ... | 77 | POINT(-47.8... | |
| 62 | *NULL* | 42 | POINT(-47.9... | |
| 63 | Sedest | 97 | POINT(-47.9... | |
| 64 | *NULL* | 202 | POINT(-47.9... | |
| 65 | *NULL* | 203 | POINT(-47.9... | |
| 66 | *NULL* | 204 | POINT(-47.9... | |
| 67 | Vila Bem Viver | 359 | POINT(-47.8... | |

Figure 5.12 – Creating a view within QGIS

Returning to our canvas with the layers from our example, be sure to arrange the layers so that the points and roads, for example, are on the top layers and not buried under polygons.

We can't gather any insights with all the layers present, as shown in *Figure 5.13*. So, now, it is time to consider a few data questions. After adding the **VIEW** property we created previously to the layers canvas, right-click and select **Zoom**:

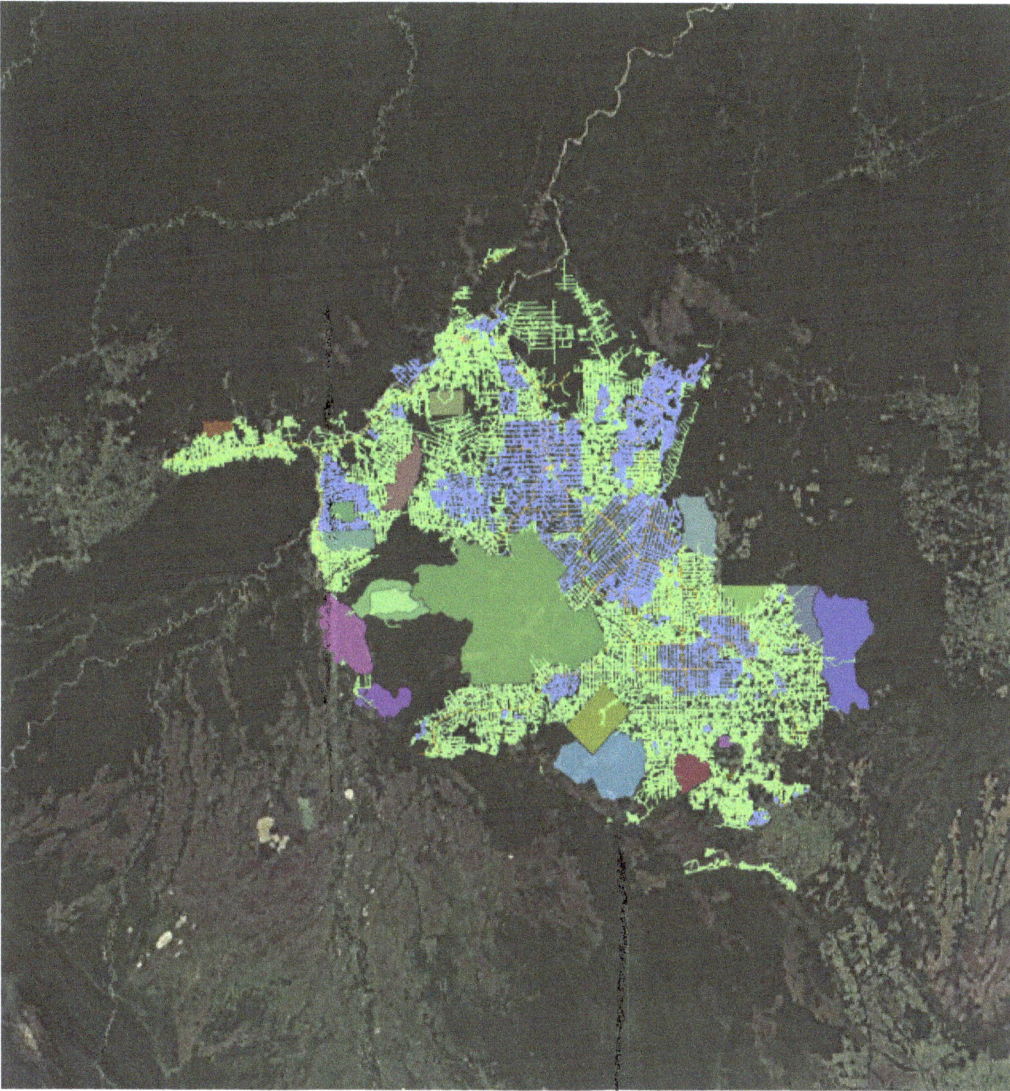

Figure 5.13 – All layers applied to the canvas

**Note**

Increasing the size of the point in **Layer Styling** increases visibility.

After zooming into the layer, as shown in *Figure 5.14*, you will now see the buildings within the protected areas, which are represented by red circles:

Figure 5.14 – Zooming into the layer to display CREATE VIEW

In *Figure 5.15*, you can see the boundary of the protected areas (beige) and the buildings located within that boundary:

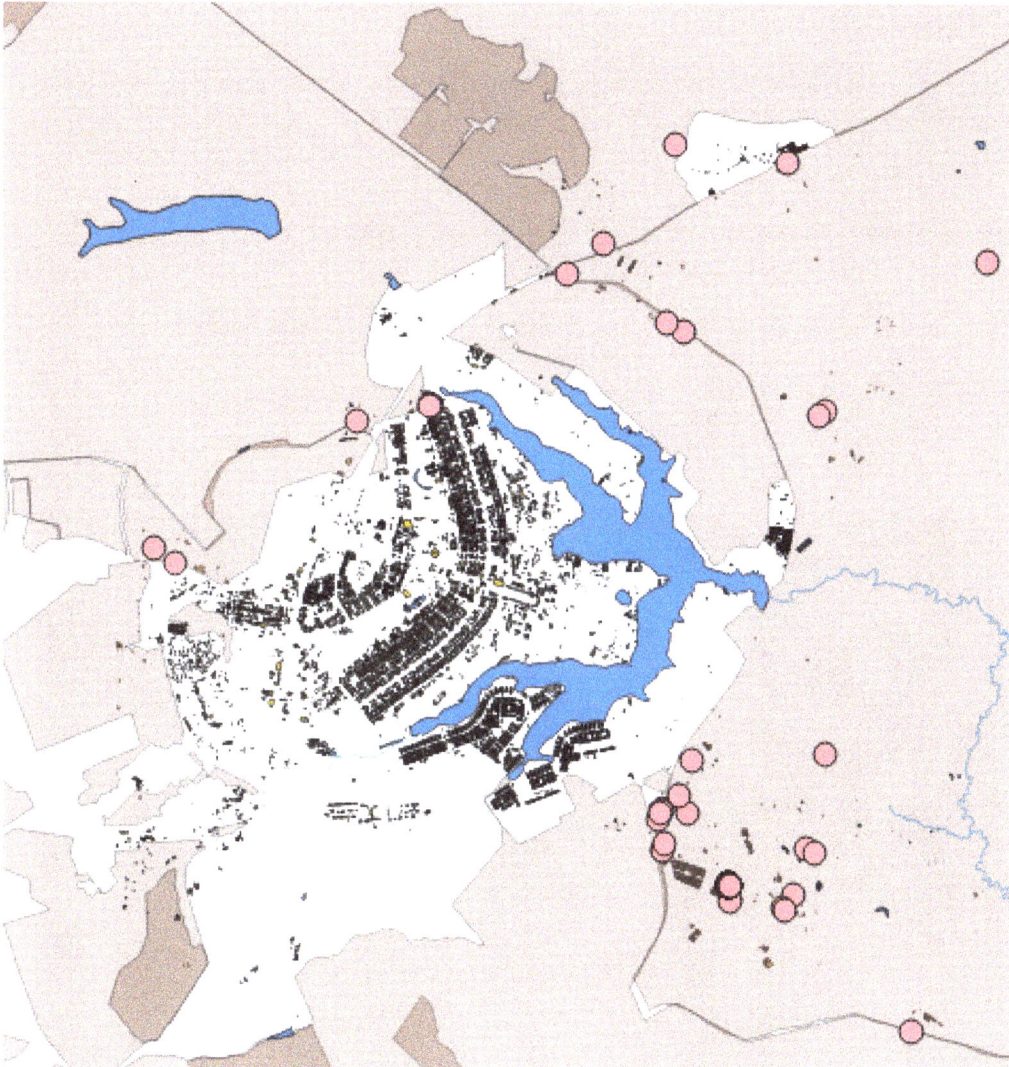

Figure 5.15 – Protected areas (beige) are shown with the buildings located within the boundaries

The water layer is styled with polygons colored blue for reference. Areas that appear brown are not necessarily the result of deforestation in a traditional sense; forests are also used for agriculture and cattle. This is not without issue or consequence, however, as you will see when we explore land use and road infrastructure.

Now that you can select the tables and data to display on the canvas, let's learn how to create a buffer. Spatial-based SQL queries for items such as containment, intersection, distances, and buffers are the foundation of spatial analysis.

## ST_Buffer

We can create buffers around specific areas in our map. This is reported as radius degrees or meters, depending on your SRID. The buffer is shown highlighted in white in *Figure 5.16*.

Here is the syntax for writing a query:

```
geometry ST_Buffer(geometry  g1, float  radius_of_buffer,
text  buffer_style_parameters = '');
```

Now, we must select the unique ID and then the parameters required by the function:

```
SELECT osm_id, ST_BUFFER(geom,.05) as geom
FROM ch5.boundary_protected_area
WHERE protection = 'reserva biológica'
```

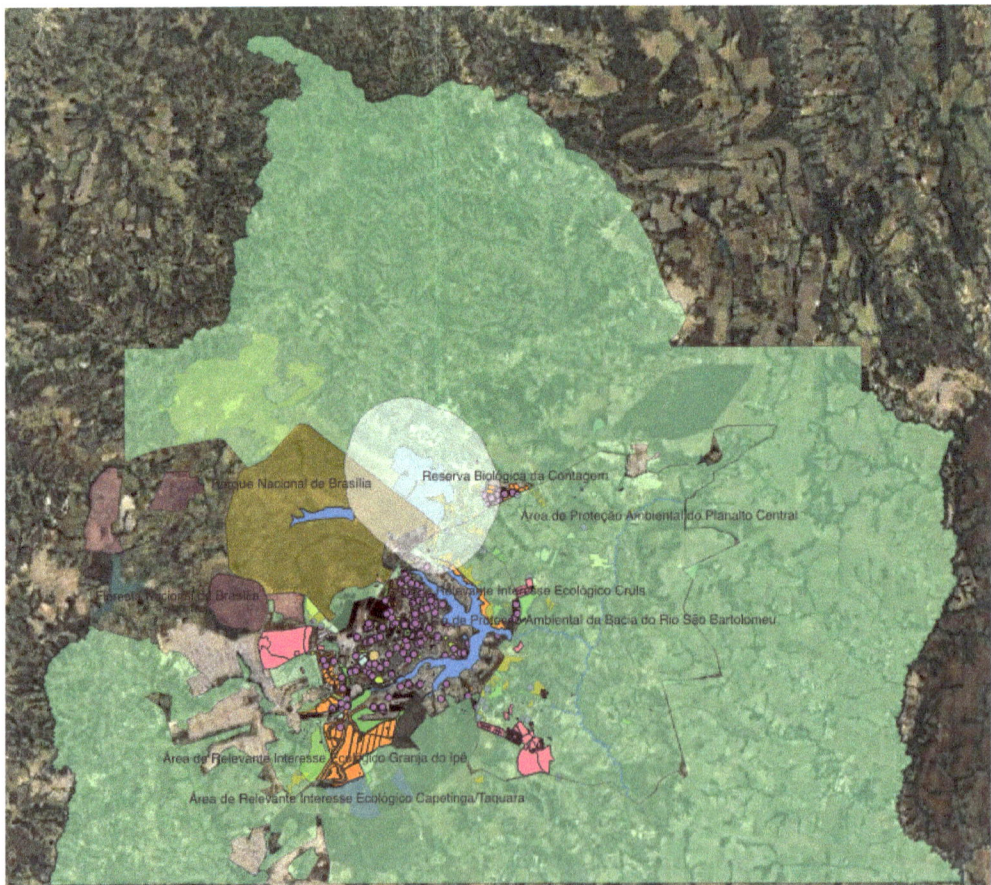

Figure 5.16 – ST_Buffer surrounding a protected area in Rondonia, Brazil

When asking data questions about specific areas on our map, we often need to select specific values in a column. In *Figure 5.15*, I have specifically requested the `protection` column (*Figure 5.16*) but need to know the identity of the values of interest – in this case, `reserva biológica`.

In addition to clicking **Layer Properties**, it is possible to access information about our data tables seen here in figure 5.17. Highlight your `Table` in pgAdmin and select **SQL** at the top of the console. Now, you can view a vertical list of column headings with information that is often helpful if you aren't seeking the detail of an attribute table.

```
6    (
7        id integer NOT NULL DEFAULT nextval('ch7.boundary_protected_area_id_seq'::regclass),
8        geom geometry(MultiPolygon,4618),
9        full_id character varying(254) COLLATE pg_catalog."default",
10       osm_id character varying(254) COLLATE pg_catalog."default",
11       osm_type character varying(254) COLLATE pg_catalog."default",
12       osm_versio character varying(254) COLLATE pg_catalog."default",
13       osm_timest character varying(254) COLLATE pg_catalog."default",
14       osm_uid character varying(254) COLLATE pg_catalog."default",
15       osm_user character varying(254) COLLATE pg_catalog."default",
16       osm_change character varying(254) COLLATE pg_catalog."default",
17       boundary character varying(254) COLLATE pg_catalog."default",
18       short_name character varying(254) COLLATE pg_catalog."default",
19       name_de character varying(254) COLLATE pg_catalog."default",
20       alt_name character varying(254) COLLATE pg_catalog."default",
21       wikipedia character varying(254) COLLATE pg_catalog."default",
22       wikidata character varying(254) COLLATE pg_catalog."default",
23       type character varying(254) COLLATE pg_catalog."default",
24       start_date character varying(254) COLLATE pg_catalog."default",
25       related_la character varying(254) COLLATE pg_catalog."default",
26       ref_cnuc character varying(254) COLLATE pg_catalog."default",
27       protection character varying(254) COLLATE pg_catalog."default",
28       protect_cl character varying(254) COLLATE pg_catalog."default",
29       operator character varying(254) COLLATE pg_catalog."default",
30       name_en character varying(254) COLLATE pg_catalog."default",
31       name character varying(254) COLLATE pg_catalog."default",
32       CONSTRAINT boundary_protected_area_pkey PRIMARY KEY (id)
33   )
```

Figure 5.17 – SQL properties in pgAdmin

When asking data questions about specific areas on your map, you often need to select specific values in a column. Asking data questions about datasets of increasing complexity often requires user-defined functions. Now, we will write our own!

## Creating functions in SQL

Creating functions can be complex but like anything else, it becomes easier with practice. Let's look through a few of the steps in detail:

1. Begin with CREATE OR REPLACE FUNCTION. This is where you name your function.

2. Next, define the function parameter(s) within parentheses.

3. Insert the RETURNS TABLE function, followed by the data type. This is a text option as the actual function will add the variable when it is run.

4. Now, select the language property of SQL as PostgreSQL since it is not limited by a single procedural language.

5. The actual query will now be included inside $$ query $$. These are called **dollar-quoted string constants ($$).**

We need to create the function (as shown in the following code) and then pass to the function what we want it to do.

First, we want to select boundary_protected_area, which is where highways intersect. We are passing text (x text), which counts as one variable. The text is entered when we run the function that was created. In our example, the text has been replaced with highway:

```
CREATE OR REPLACE FUNCTION getprotecteda (x text)
RETURNS TABLE("geom" geometry,"boundary_protected_area" text)
AS $$
SELECT boundary_protected_area.geom,ch5.boundary_protected_
area.osm_id
FROM ch5.boundary_protected_area,ch5.highway
WHERE st_Intersects (ch5.boundary_protected_area.geom,ch5.
highway.geom) AND ch5.highway.highway IS NOT NULL ;
$$ LANGUAGE SQL;
```

We must run the following function to get our output:

```
SELECT * FROM getprotecteda ('highway')
```

Visualizing the output in *figure 5.18* in QGIS:

Figure 5.18 – Running the function to generate the output

We can now view the region of interest and observe the dense network of highways running through the protected area, as shown in *Figure 5.19*:

Figure 5.19 – Protected region and proximity to highway networks

Returning to pgAdmin, you can also select a particular boundary and explore its specific relationships. We will randomly select osm_id '10160475'. You can see the boundaries in the **Geometry Viewer** area in *Figure 5.20*:

```
CREATE OR REPLACE FUNCTION getprotecteda (x text)
RETURNS TABLE("geom" geometry,"boundary_protected_area" text)
AS $$
SELECT boundary_protected_area.geom,ch5.boundary_protected_
area.osm_id
FROM ch5.boundary_protected_area,ch5.highway
WHERE st_Intersects (ch5.boundary_protected_area.geom,ch5.
highway.geom) AND ch5.highway.highway IS NOT NULL ;
$$ LANGUAGE SQL;
SELECT * FROM getprotecteda('10160475')
```

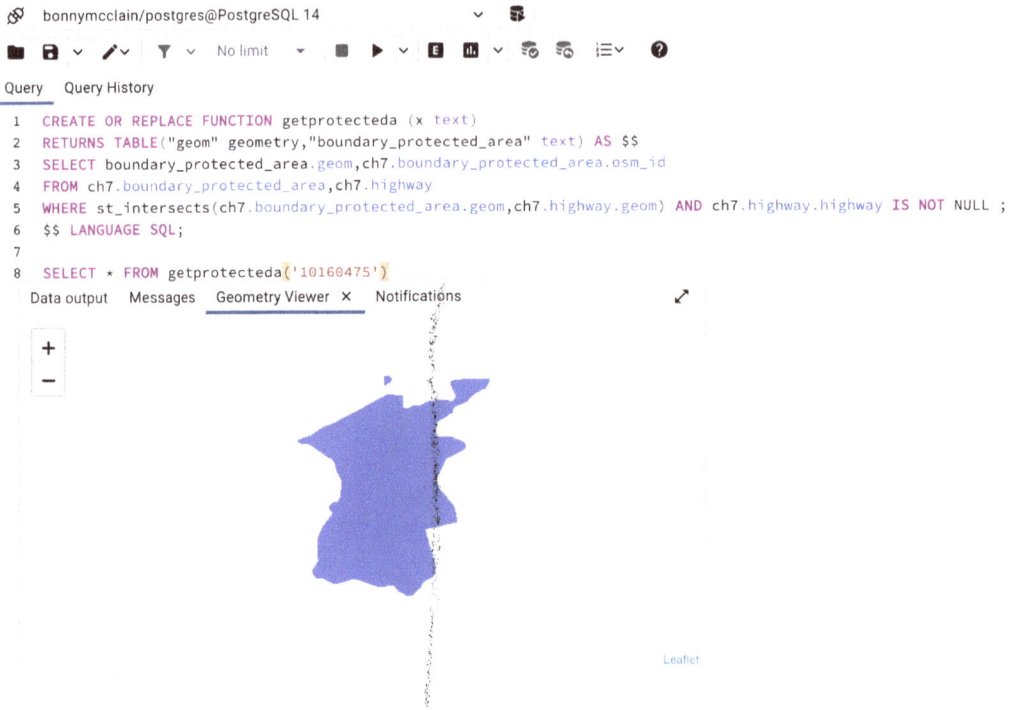

Figure 5.20 – Creating a function to view protected areas

Running the same code in QGIS provides the map scene shown in *Figure 5.21*. To update the layer that's already been loaded onto the canvas, we need to right-click on the query layer and select **update SQL Layer**. You have a wide range of base maps to select from in QGIS:

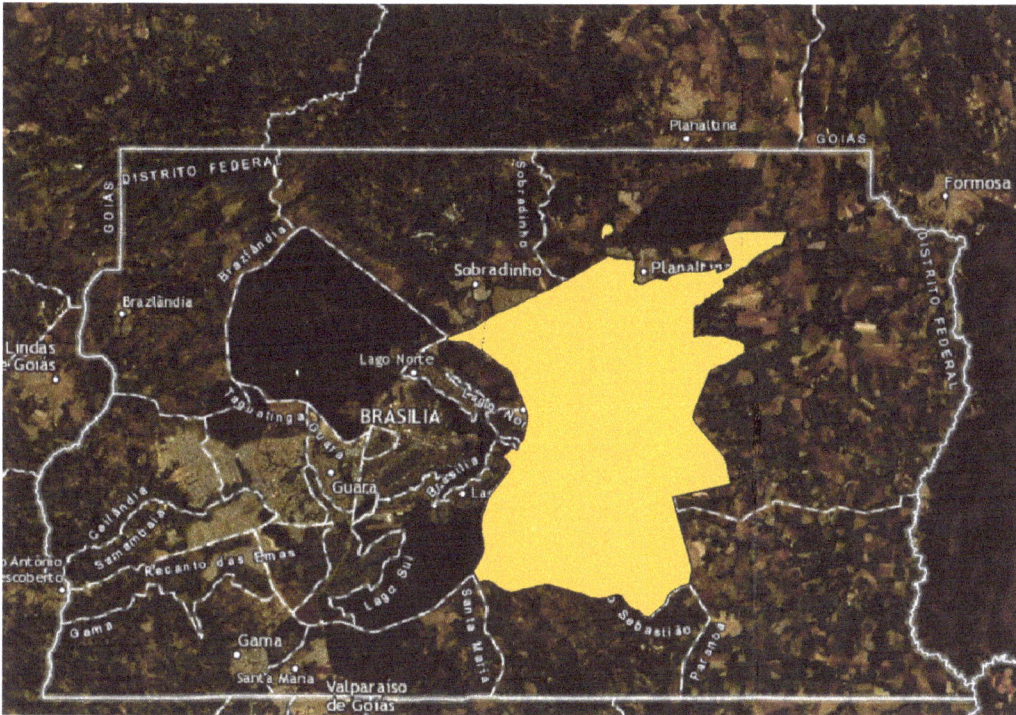

Figure 5.21 – Viewing the boundaries in the getprotecteda function

The query runs under the hood in QGIS when adding the updated layer to the canvas, as shown in the following SQL statement generated by QGIS:

```
SELECT row_number() over () AS _uid_,* FROM (SELECT * FROM
getprotecteda('10160475')
) AS _subq_1_
```

Now, we can see the updated layer for the selected OSM-ID in the region with the associated intersection of the network of the highway (*Figure 5.22*):

Figure 5.22 – Updated SQL layer showing the region of interest with intersecting highways

Next, let's move on to mining areas in Brazil.

## Mining areas in Brazil

If we wish to explore the region by area, we can use the ST_Area SQL function.

### ST_Area

ST_Area will return the area of the geometry, but you need to pay attention to the SRID to determine the units:

```
float ST_Area(geometry g1);
```

To explore the mining areas in Brazil, we can use the ST_Area function. In pgAdmin, we can observe these results in **Geometry Viewer**:

```
SELECT
geom ST_Area(geom)/10000 AS hectares
FROM ch5.mining_polygons
WHERE ch5.mining_polygons.country_name = 'Brazil'
```

The following *figure 5.23* shows the results of our query:

Query    Query History

```
1    SELECT
2    geom, ST_Area(geom)/10000 AS hectares
3    FROM ch5.mining_polygons
4    WHERE ch5.mining_polygons.country_name = 'Brazil'
5
```

Data output    Messages    Geometry Viewer  ✕    Notifications

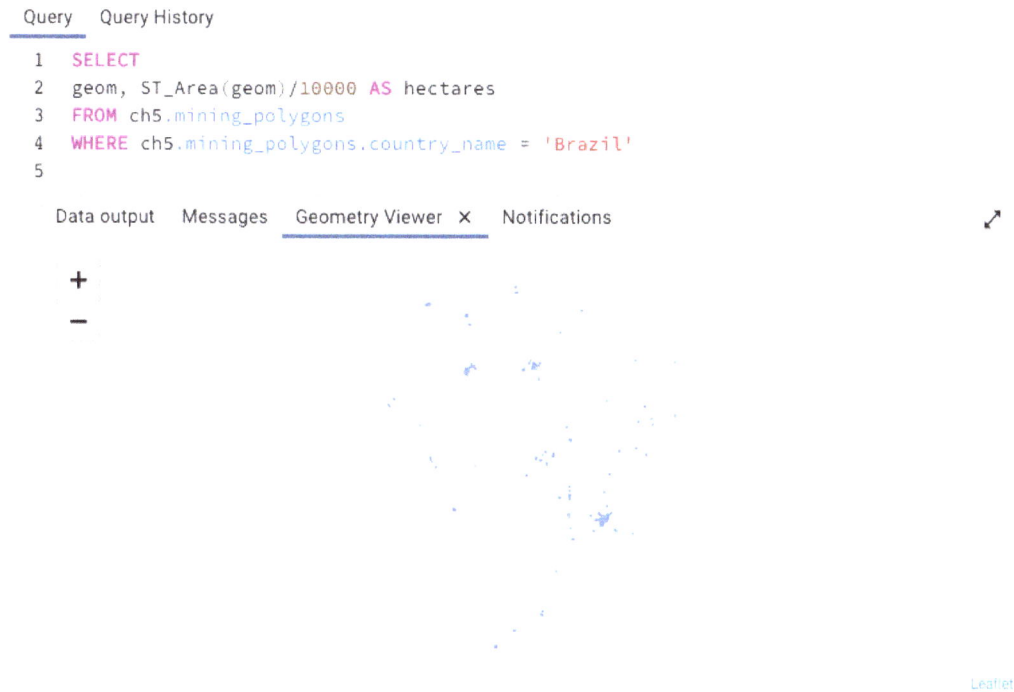

Figure 5.23 – Polygons representing mining areas in Brazil

Becoming familiar with this function allows you to explore the area occupied by indigenous populations shown in *Figure 5.24* by running the following code:

```
SELECT
    nm_uf,
    ST_Area(geom)/10000 AS hectares
FROM ch5."br_demog_indig_2019"
ORDER BY hectares DESC
LIMIT 25;
```

Query    Query History

```
1   SELECT
2     geom,nm_uf,
3     ST_Area(geom)/10000 AS hectares
4   FROM ch5."br_demog_indig_2019"
5   ORDER BY hectares DESC
6
7
8
```

Data output    Messages    Geometry Viewer ✕    Notifications

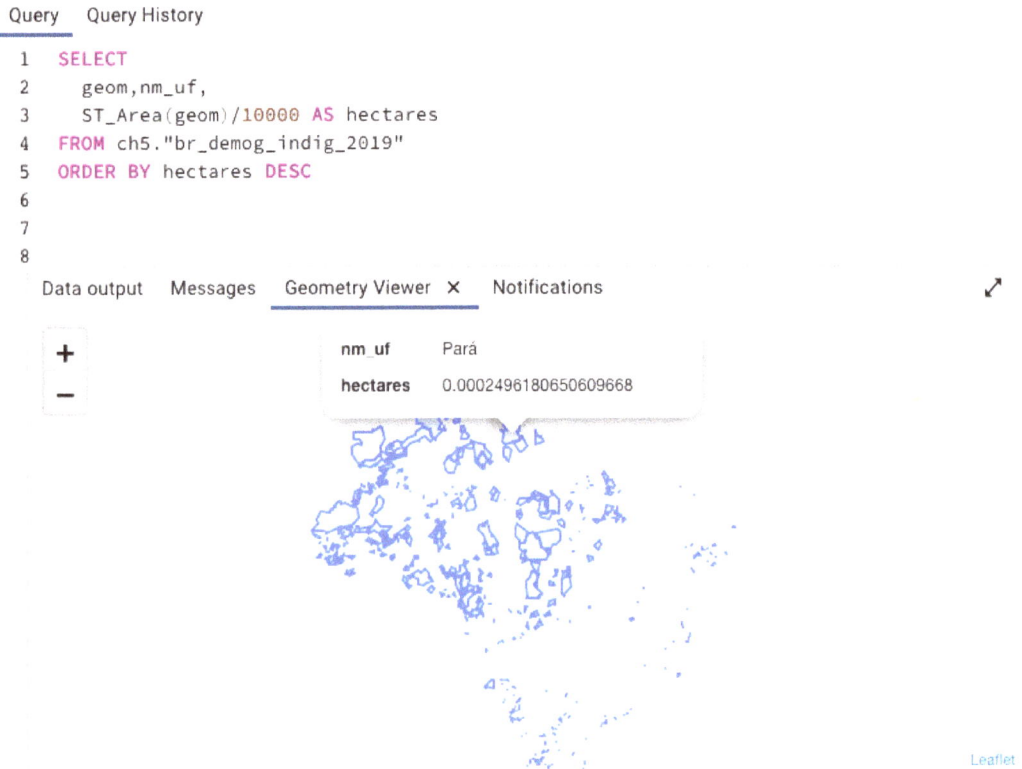

| nm_uf | Pará |
|-------|------|
| hectares | 0.0002496180650609668 |

Leaflet

Figure 5.24 – Areas occupied by indigenous populations

Next, let's move on to ST_DWITHIN.

## ST_DWITHIN

This function returns true if the requested geometries are within a given distance. For example, in the following code, a request has been made for roads within 300 meters of a deforested area, as defined by "double precision distance_of_srid":

```
boolean ST_DWithin(geometry g1, geometry g2, double precision
distance_of_srid);
```

The distance is provided in the units defined by the spatial reference system of the geometries:

```
SELECT DISTINCT ON (ch5."Roads".objectid) ch5."Roads".objectid,
ch5."Roads".status,ch5."Roads".geom, ch5."Deforested_Area".
objectid
FROM ch5."Roads"
```

```
LEFT JOIN ch5"Deforested_Area" ON ST_DWITHIN(ch5."Deforested_
Area".geom, ch5."Roads".geom, 300)
ORDER BY ch5."Roads".objectid, ST_Distance(ch5."Roads".geom,
ch5."Deforested_Area".geom);
```

I tend to run queries in pgAdmin and save them there. In *Figure 5.25*, we can see a partial rendering of the data. I only requested a minimum number of rows to expedite the runtime since I will head over to QGIS to see the visualization:

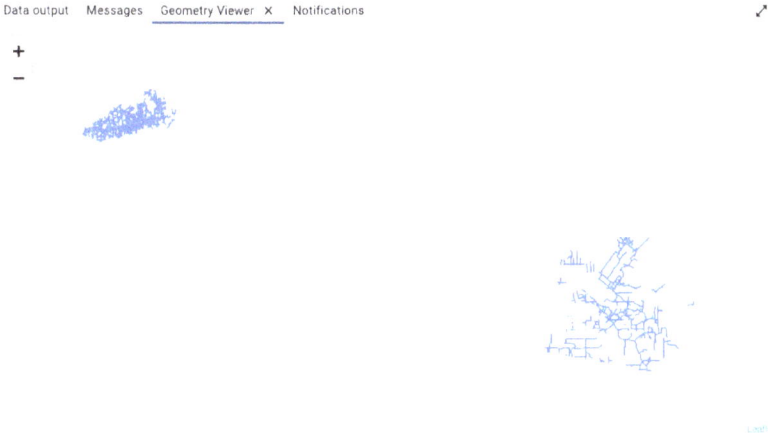

Figure 5.25 – Exploring the data in Geometry Viewer in pgAdmin

When observing the data, I noticed that there were several classifications of roads. If you right-click on the road layer, you will see the option to filter. We can also do this with SQL code, but it is only a few clicks away within the QGIS console, as shown in *Figure 5.26*:

1.  Select the Field property you would like to filter. In this example, this will be status.

2.  If you are not familiar with the variable, double-click it; options will appear in the **Values** window on the right.

3.  Select a **Sample** if you want to have a quicker response.

    I am filtering on only the official roads.

4.  Clicking on the **Field** and **Value** properties will add them to the **Expression** window at the bottom of the console.

5.  Select **Test** to see if your expression is valid. You can see the results of my query and that it was successful, returning 613 rows.

6.    Select **OK**; you will see the filter icon next to the query in your **Layer** panel:

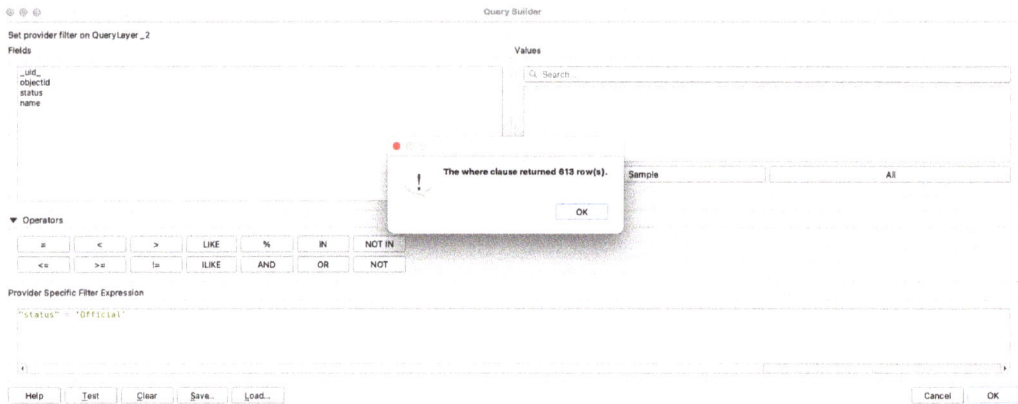

Figure 5.26 – Filtering the roads field so that it only includes official roads

The roads we filtered are now shown in pink in *Figure 5.27*. We can now add the layers to the canvas and observe the results of the query. You can remove the fill in Layer Stying and retain the outline of the polygon. We are observing the proximity of roads to the indigenous territories in *Figure 5.27*. But the addition of the deforested areas is where we can begin to see the impact of the roads on the surrounding areas, as observed by the shaded area in *Figure 5.28*:

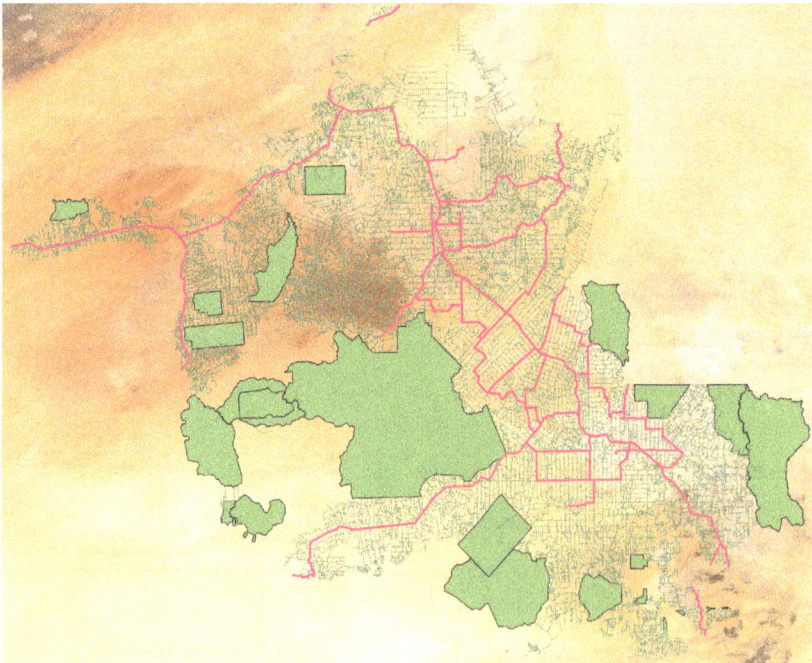

Figure 5.27 – Roads and their relationship with indigenous territories in green polygons

Upon observing the impact of building the road infrastructure on deforestation in *Figure 5.28*, we can see it is related to a decrease in vegetation, but we will explore other ways to measure the impact in the next chapter as we expand our skills and ask bigger questions:

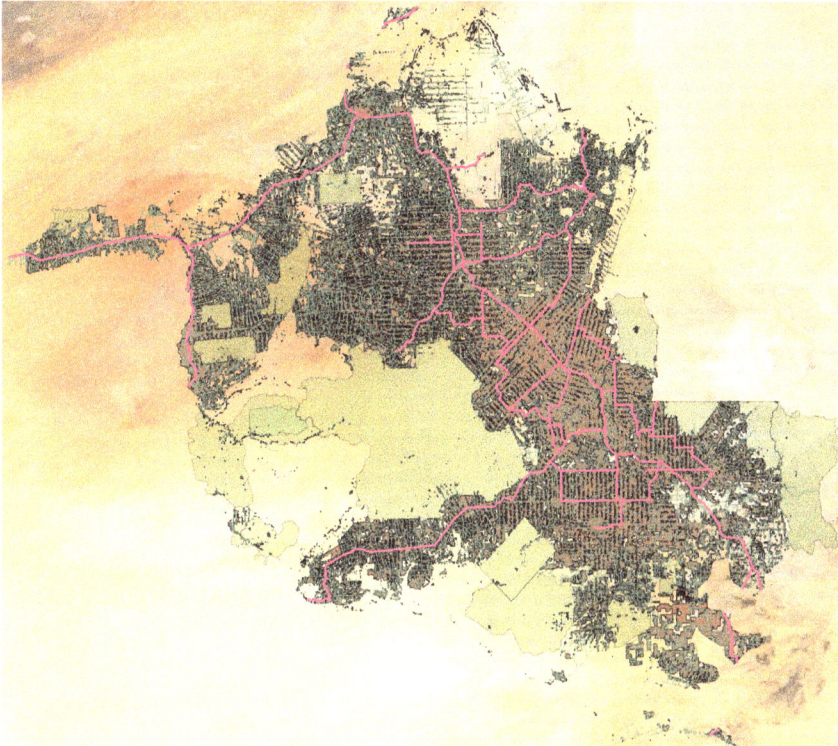

Figure 5.28 – Deforested areas added to the canvas in QGIS

The ability to bring different layers into our queries highlights the power of combining QGIS and SQL into our emerging data questions.

Now, let's summarize everything you have learned.

## Summary

In this chapter, we prepared data so that it could be imported and analyzed. You were introduced to new spatial functions so that you can estimate areas of polygons and define relationships between and within polygons. You even learned how to create your own functions. These are handy to know especially when you might return to datasets that are frequently updated.

In the next chapter, we will expand on building SQL queries in the QGIS graphical query builder that you were introduced to when filtering the roads data. SQL is indeed a powerful tool for creating efficient workflows and optimizing the power of working with GIS tools such as QGIS.

# Building SQL Queries Visually in a Graphical Query Builder

In the previous chapter, you were introduced to topological spatial queries such as distance, buffers, and proximity to characteristics of a variety of entities.

In this chapter, we will continue exploring the tools and concepts available for learning about SQL and applying this knowledge to geospatial analysis. Writing SQL queries into **QGIS** allows you to visualize output directly on a map. pgAdmin also allows a quick look at a snapshot of your data using **Geometry Viewer** but without a lot of customization.

Discovering quick and efficient tools for integrating database queries with the SQL query builder builds on the query language you are already scripting.

In this chapter, you will learn about the following topics:

- How to access the **Graphical Query Builder (GQB)**
- How to write complex queries in the SQL query builder
- Beginning to build advanced frameworks for our data exploration

## Technical requirements

The data for the exercises in this chapter can be found on GitHub at `https://github.com/PacktPublishing/Geospatial-Analysis-with-SQL`

The following datasets will be used in this chapter:

- `Airport_Influence_area`
- `Census_Designated_places_2020`

- `Income_per_Capita` per census tract
- American Community Survey (ACS) – 1 year data

As we learned about geometric and geographic spatial queries in defining coordinates and locations, we investigated points, lines, and areas of polygons. When referring to a topological query, we are mainly interested in where objects are located and where they might intersect, in addition to specific attributes.

Topological SQL queries investigate spatial relations between tables or entities within a database. Topological criteria focus on the arrangement of entities in space. These arrangements are preserved regardless of projection. We have already explored spatial relationships such as distance, intersections, and geometric relationships, for example. We can calculate characteristics and relationships as well as create new objects.

The characteristics of topological queries are qualitative and examined over two-dimensional space, essentially exploring region boundaries to define a relation or dependency. In practice, you can think of these queries as asking questions about adjacency, connectivity, inclusion, and intersection. You may have heard of non-topographic data being referred to as *spaghetti data*. In fact, ESRI shapefiles are representative of this type of data. Shapefiles store vector data and build connections on top of the feature class. Topographical data, by contrast, stores relationships such as nodes or points and polygons capturing shared boundaries and adjacency to avoid overlapping polygons or lines that omit connectivity to other lines.

## Conditional queries

We evaluate these relations with boolean test queries (`true` or `false`), numerical distance queries, or action queries by testing relationships based on geometries. These are all conditional queries. For example, any of the functions listed can be inserted into this sample query structure. We are looking at the relationship between two different geometries in a single table. We will expand this to multiple tables, but as an introduction, this is the framework:

```
SELECT FUNCTION (geom1, geom2 FROM name_table)
```

These are the sample functions that evaluate the function and return data if it is `true`.

The following boolean functions are either true or `false` based on the relationships you query:

FUNCTIONS:

Equals(),

Disjoint(),

Touches(),

Within(),

Overlaps(),

Crosses(),

Intersects (),

Contains(),

Relates()

You will create a QGIS workflow for exploring datasets, but first, let's get familiar with the interface.

# SQL data queries using the graphical query builder

The query builder is SQL-like in its language syntax, and it is worth an introduction as part of an independent workflow. For example, simply because I code in Python or write SQL queries doesn't mean I avoid a good shortcut here and there. Let's take a look at the query builder and then explore how to integrate SQL syntax into our filters and code expressions.

First, we will access the query builder. Follow these steps:

1.  Select **Layer** from the menu in QGIS, scroll down to **Layer Properties**, and activate it by clicking on it, as shown in *Figure 6.1*:

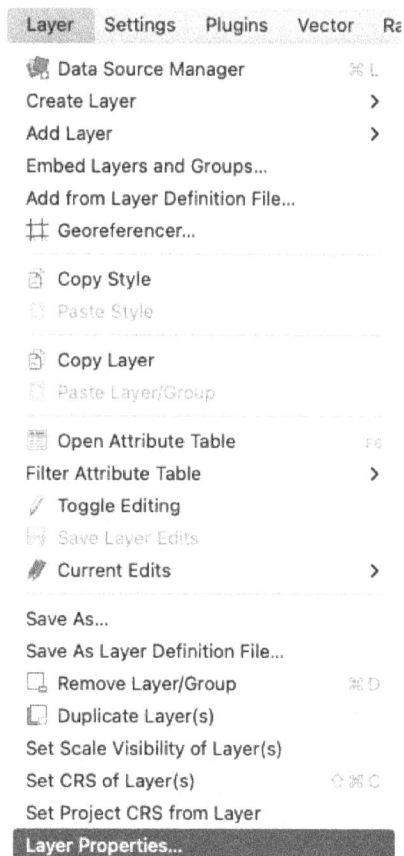

Figure 6.1 – Layer properties in QGIS

2.  When you click on the **Layer Properties** option, the window opens, as shown in *Figure 6.2*. Select **Source** and the page will display the following options:

    - **Settings**: Displays the layer name and data source encoding. The default is UTF-8.

    - **Assigned Coordinate Reference System (CRS)**: This operates as a fix in case the wrong projection is displayed. If you need to change the projection beyond this instance of the project, you will need to go to the **Processing Toolbox** option and select the **Reproject** layer.

    - **Geometry**: These options are typically defined already but if not, you can modify them here.

3.  Next, select the **Provider Feature Filter** option, and the **Query Builder** button will appear in the lower-right corner of the window. Click on **Query Builder** and you will see a window displayed, as in *Figure 6.2*. The fields that populate will be from the layer you selected in the **Source** window:

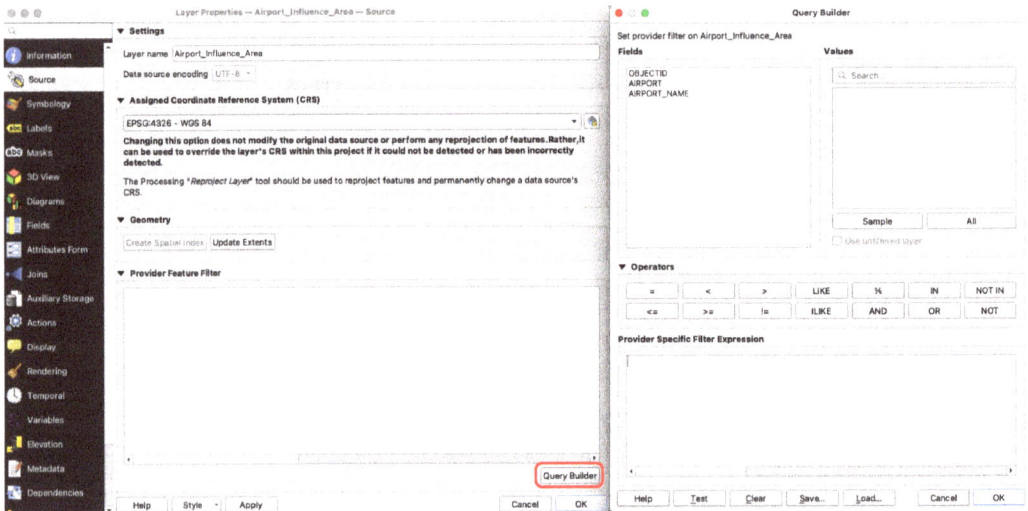

Figure 6.2 – Query Builder in QGIS

The Airport_Influence_Area data layer is from the *Open Data* portal for Los Angeles County. The polygons denote an airport influence area within 2 miles of an airport, representing noise, vibration, odors, or other inconveniences unique to the proximity of an airport. According to a state's Division of Aeronautics, the *airport influence area* is determined by an airport landuse commission and varies for each airport.

In this quick introductory example, you can choose AIRPORT_NAME and either see a sample of the values in the right window or select **All** to see all values. Alternatively, if you are interested in a specific airport, you can search for it in the **Search** window. In addition, simply selecting **Fields** and hitting **Test** will show you how many rows are in your data. I often start here, write the query, and then look to see whether the number of rows provided after the query makes sense.

*Figure 6.3* includes a simple query where we are selecting AIRPORT_NAME and the value as **Burbank**. By clicking on **Fields** and adding an operator and a value, the canvas will update with your filter. The **Test** button allows you to run the query to see whether it is valid:

Figure 6.3 – A sample filter expression in query builder in QGIS

We expect a single row since we are requesting a single airport. Other fields from different layers will have multiple value options and will return the valid rows for your query. For example, let's look at Income_per_Capita_ (Census_tract) in Los Angeles County. In *Figure 6.4*, we can see the styling layer by graduated equal count intervals:

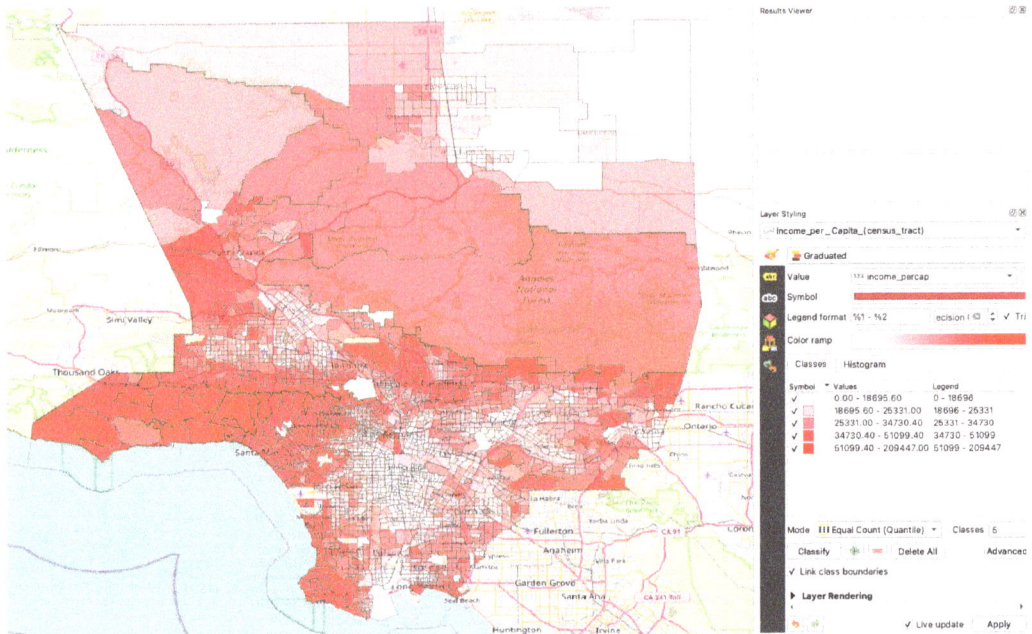

Figure 6.4 – Income per capita in census tracts in Los Angeles County (QGIS)

Returning to the query builder, we need to request an actual value. We can request values using operators.

How about census tracts where income per capita is less than $31,602?

The operators are, = (equal), < (less than), > (greater than), <= (less than or equal to), >= (greater than or equal to), != (not equal, and % (wildcard).

The remaining reserved words are self-explanatory, with the exception of ILIKE, which indicates insensitive to the case. The last two options, for example, might be useful if you are looking for cities that begin with a letter; that is, returning all values that begin with b would use b%. The wildcard (%) moves to the first place when looking for words that end with the indicated letter, %b.

Using the query builder, we select the AIRPORT field and all values that begin with b, regardless of case (ILIKE). The clause returns two rows, as displayed in *Figure 6.5*:

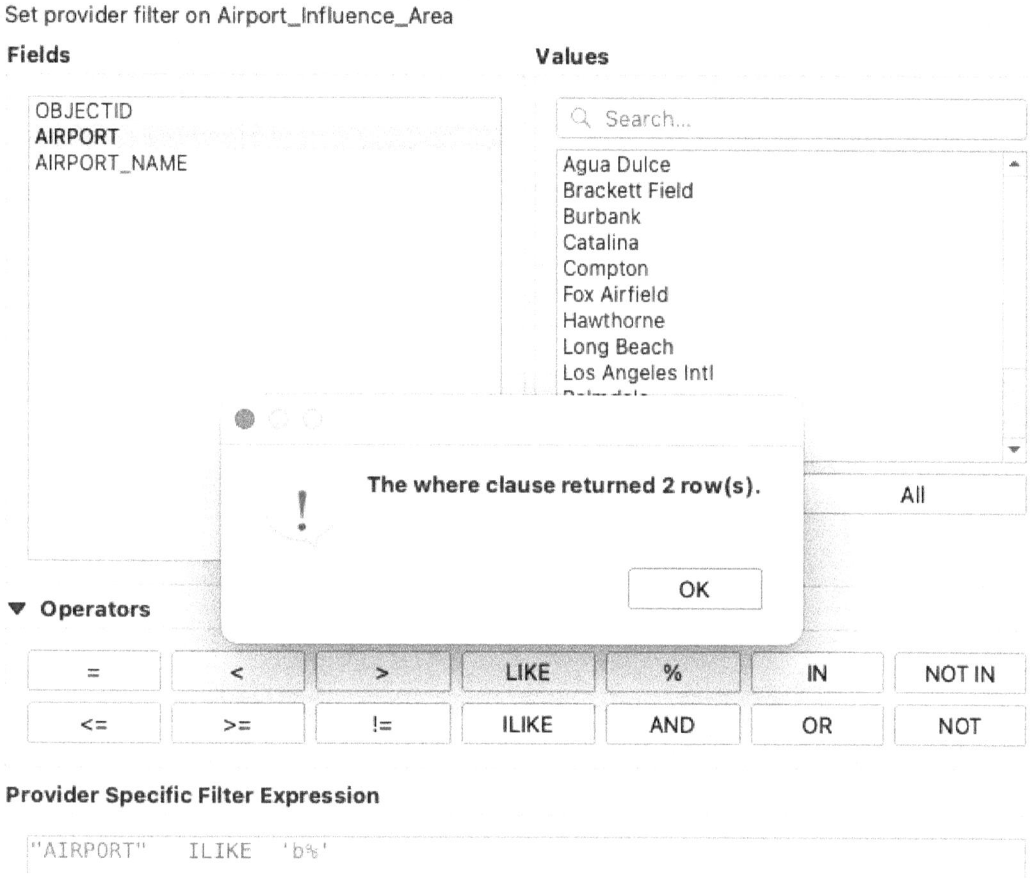

Set provider filter on Airport_Influence_Area

**Fields**

OBJECTID
**AIRPORT**
AIRPORT_NAME

**Values**

🔍 Search...

Agua Dulce
Brackett Field
Burbank
Catalina
Compton
Fox Airfield
Hawthorne
Long Beach
Los Angeles Intl

**The where clause returned 2 row(s).**

All

OK

▼ **Operators**

| = | < | > | LIKE | % | IN | NOT IN |
|---|---|---|------|---|----|----|
| <= | >= | != | ILIKE | AND | OR | NOT |

**Provider Specific Filter Expression**

```
"AIRPORT"  ILIKE  'b%'
```

Figure 6.5 – Using the ILIKE operator in QGIS to select airports

Notice that in *Figure 6.6*, without the ILIKE operator the clause returns 0 rows because the values are returned capitalized, and we requested the values with a lowercase b:

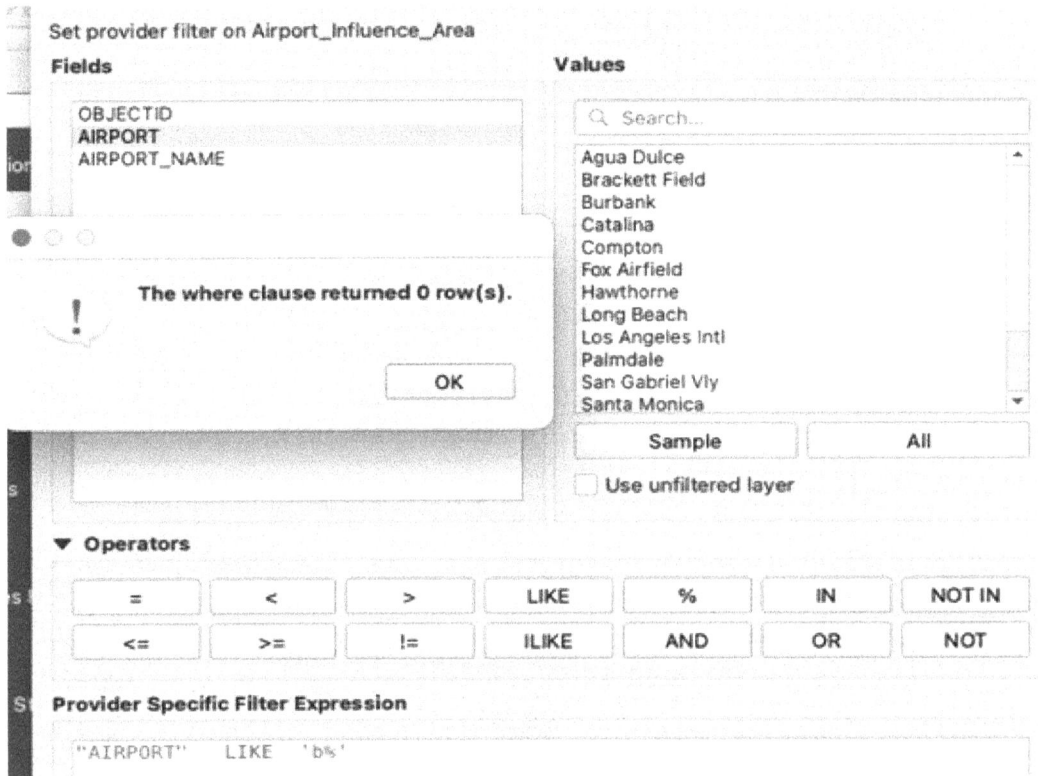

Figure 6.6 – Selecting all airports beginning with the letter b using the LIKE operator in QGIS

These simple queries are a nice introduction to how queries are structured. Conditional queries have a filter expression that evaluates a relationship and returns the output. These examples are looking at a single table and the data values that are included. In the next section, we are going to begin working across tables.

# Aggregation and sorting queries

These quick queries are useful, but often, you will need to manage more complex queries or access more than one table in your database.

## Understanding census-designated places

Census-designated places are statistical entities used by the US Census Bureau. Think of unincorporated places with boundaries outside of incorporated places but relying on services provided to the general area. They are basically communities that look like cities.

We will start with something more interesting than locating an airport with b included in its name. Let's use the **database manager** query manager to write a simple SQL query. Select all the rows from our table. You can indicate an alias if you want to simplify the code, `SELECT * FROM ch6."Census_Designated_Places_2020" c`, and now, any time you refer to the table, it can simply be `c.geom` or `c.name`, for example. I will start doing this practice as we move to complex queries but I find it can be confusing in teaching environments, so I will continue to write out the name of the tables; however, you can certainly explore using the alias of your choosing.

Write the following SQL query and click **Execute**:

```
SELECT * FROM ch6."Census_Designated_Places_2020"
```

The table will load with the results of the query, as shown in *Figure 6.7*:

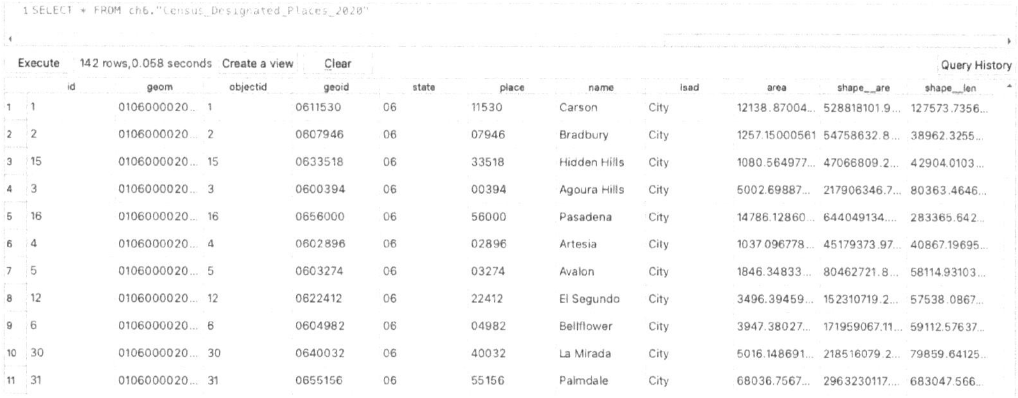

```
1 SELECT * FROM ch6."Census_Designated_Places_2020"
```

| | id | geom | objectid | geoid | state | place | name | lsad | area | shape__are | shape__len |
|---|---|---|---|---|---|---|---|---|---|---|---|
| | | | | | | | | | | | Query History |
| 1 | 1 | 0106000020... 1 | | 0611530 | 06 | 11530 | Carson | City | 12138.87004... | 528818101.9... | 127573.7356... |
| 2 | 2 | 0106000020... 2 | | 0807946 | 06 | 07946 | Bradbury | City | 1257.15000561 | 54758632.8... | 38962.3255... |
| 3 | 15 | 0106000020... 15 | | 0633518 | 06 | 33518 | Hidden Hills | City | 1080.564977... | 47066809.2... | 42904.0103... |
| 4 | 3 | 0106000020... 3 | | 0600394 | 06 | 00394 | Agoura Hills | City | 5002.69887... | 217906346.7... | 80363.4646... |
| 5 | 16 | 0106000020... 16 | | 0656000 | 06 | 56000 | Pasadena | City | 14786.12860... | 644049134.... | 283365.642... |
| 6 | 4 | 0106000020... 4 | | 0602896 | 06 | 02896 | Artesia | City | 1037.096778... | 45179373.97... | 40867.19695... |
| 7 | 5 | 0106000020... 5 | | 0603274 | 06 | 03274 | Avalon | City | 1846.34833... | 80462721.8... | 58114.93103... |
| 8 | 12 | 0106000020... 12 | | 0622412 | 06 | 22412 | El Segundo | City | 3496.39459... | 152310719.2... | 57538.0867... |
| 9 | 6 | 0106000020... 6 | | 0604982 | 06 | 04982 | Bellflower | City | 3947.38027... | 171959067.11... | 59112.57637... |
| 10 | 30 | 0106000020... 30 | | 0640032 | 06 | 40032 | La Mirada | City | 5016.148691... | 218516079.2... | 79859.64125... |
| 11 | 31 | 0106000020... 31 | | 0655156 | 06 | 55156 | Palmdale | City | 68036.7567... | 2963230117... | 683047.566... |

Execute   142 rows,0.058 seconds   Create a view   Clear

Figure 6.7 – Selecting a table in the query window: QGIS

Selecting all columns in your dataset is an easy way to look over the data and the type of data. Let's select a WHERE statement and find the **El Segundo** location. I grabbed the city randomly, and you can do the same or pick a different location. El Segundo is the epicenter for Los Angeles sports teams, the future location for the 2028 Olympics, and home of the *Los Angeles Times*, to name a few bits of history:

```
SELECT * FROM ch6."Census_Designated_Places_2020" WHERE name =
'El Segundo'
```

*Figure 6.8* shows the city and area of El Segundo. Now, let's add it to our map:

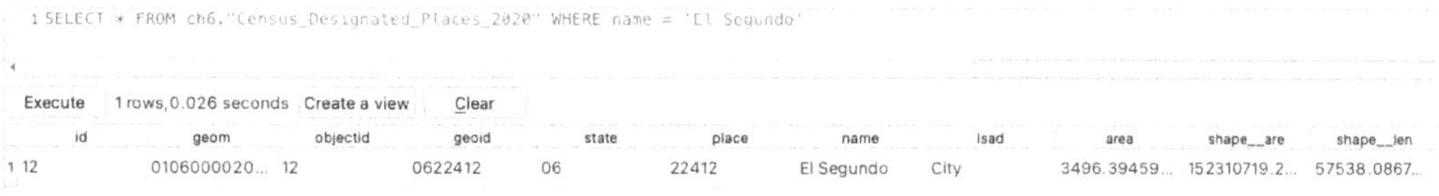

```
1 SELECT * FROM ch6."Census_Designated_Places_2020" WHERE name = 'El Segundo'
```

Execute   1 rows,0.026 seconds   Create a view   Clear

| | id | geom | objectid | geoid | state | place | name | lsad | area | shape__are | shape__len |
|---|---|---|---|---|---|---|---|---|---|---|---|
| 1 | 12 | 0106000020... 12 | | 0622412 | 06 | 22412 | El Segundo | City | 3496.39459... | 152310719.2... | 57538.0867... |

Figure 6.8 – Result of the QGIS query builder

Scroll down the page and check the **Load as new layer** checkbox. You also have the opportunity to label the new layer in the **Layer name (prefix)** panel, shown in *Figure 6.9*. Load the data and wait for `QueryLayer` to become visible in the Layer panel. Make sure you move it to the top of the layers so that it won't be hidden. If you don't choose a name **Query** layer will be the default:

| | id | geom | objectid | geoid | state | place | name | lsad | area | shape__are | shape__len | Query History |
|---|---|---|---|---|---|---|---|---|---|---|---|---|
| 1 | 12 | 0106000020... | 12 | 0622412 | 06 | 22412 | El Segundo | City | 3496.39459... | 152310719.2... | 57538.0867... | |

✓ Load as new layer

Column(s) with unique values   id          ▼  ✓ Geometry column   geom          ▼     Retrieve columns

Layer name (prefix)          Set filter

Avoid selecting by feature id          Load

Cancel

Figure 6.9 – Loading the new layer onto the canvas in QGIS

To see the updated canvas, right-click on Query Layer and select **Update SQL Layer...**, as shown in *Figure 6.10*:

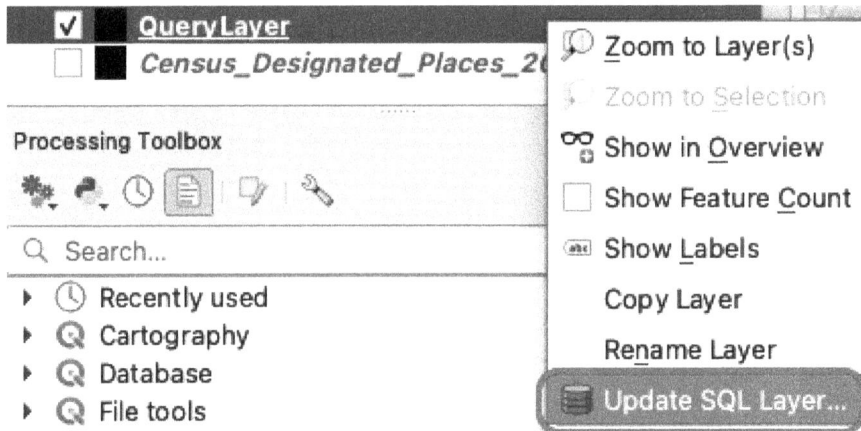

Figure 6.10 – Updating the SQL layer in QGIS

*Figure 6.11* shows the updated SQL layer, creating an additional column that you can customize or leave with the default _uid_ value, an abbreviation for **unique identifier**:

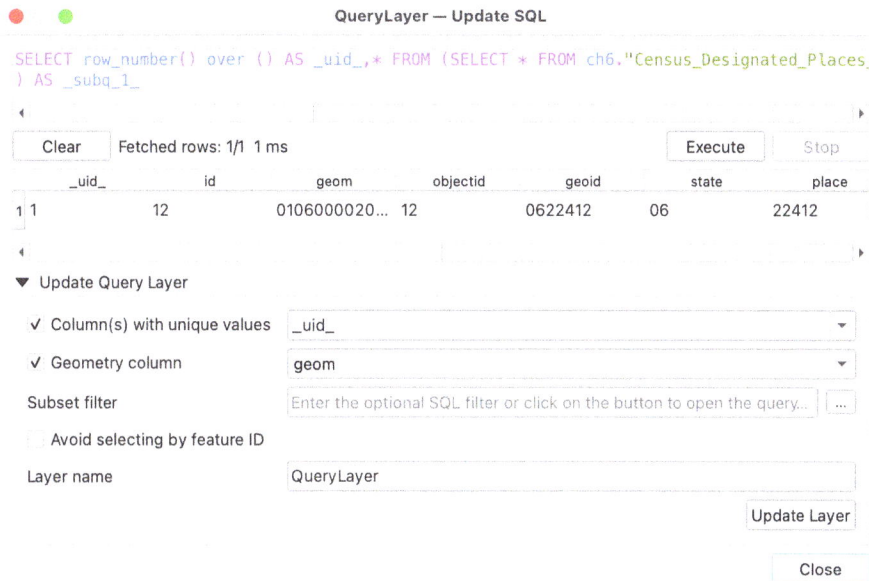

Figure 6.11 – Updating the layer with the updated SQL query

The green polygon in *Figure 6.12* is the El Segundo census place, part of unincorporated Los Angeles County. The layer for income shows light green at the **Lowest Income per Capita By** census tract surrounded by the highest census tract income levels in a darker red color:

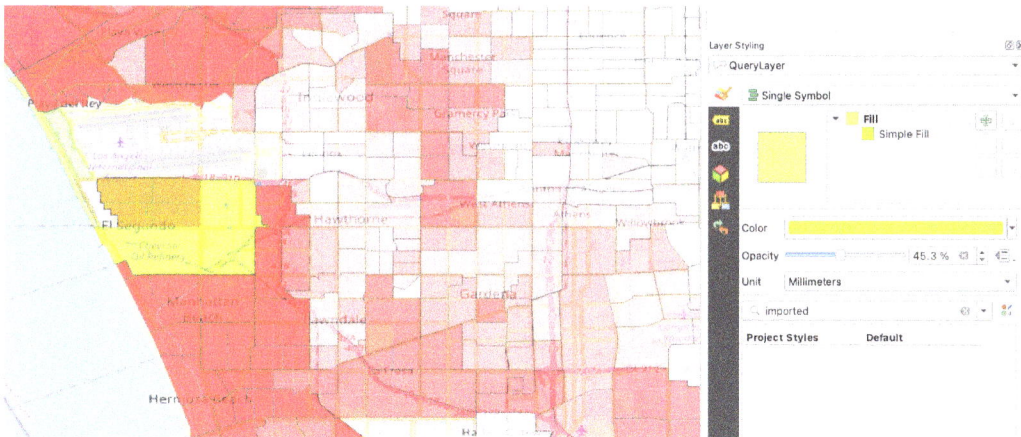

Figure 6.12 – The updated canvas in the QGIS query builder

Before we build a more complex story, let's become familiar with the SQL query builder. It's similar to the query builder we have been using, but now we have the flexibility of writing SQL queries and customizing our data questions.

I recommend adding **DB Manager** to your toolbar if you have not done so already. You can find it in the **View** dropdown in the menu bar. Scroll down to **Toolbars** and click on **Database Toolbar**—**DB Manager** will open.

## The SQL Query Builder

The **SQL Query Tool option** available under DB Manager is also available for complex queries and filtering columns. Notice the little SQL icon in the upper-left corner of the query window. Click the icon, and the new SQL Query Builder will pop up.

*Figure 6.13* shows us the different areas where we can select data we are interested in from the relevant tables, write a WHERE query, group our results by whichever variable we choose, and load the results as a new layer in our layer panel.

There are a host of other options on the lower-right panel that we will explore in more detail in *Chapter 7, Exploring PostGIS for Geographic Analysis*:

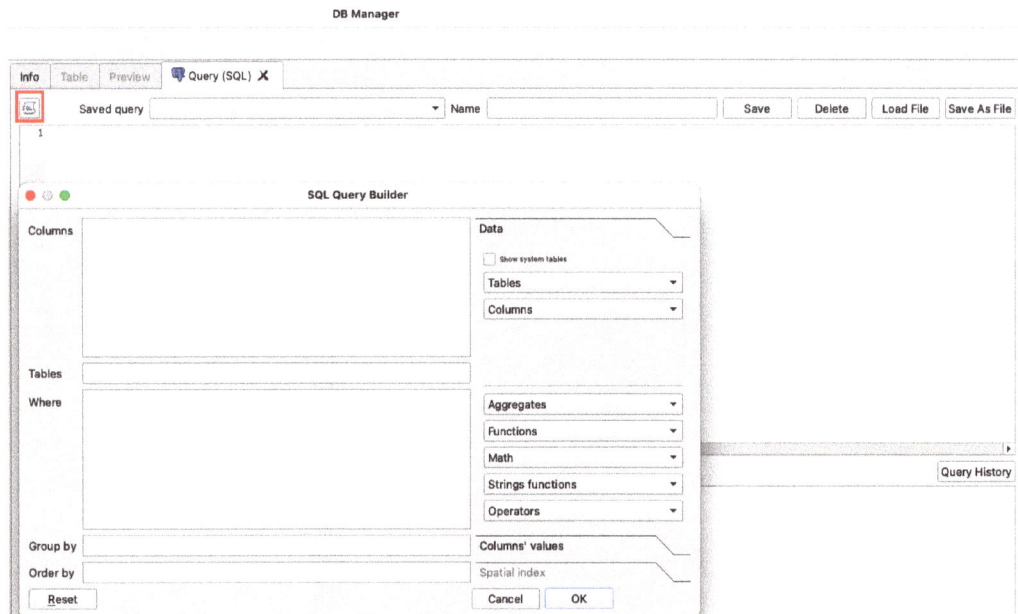

Figure 6.13 – DB Manager SQL Query Builder: QGIS

Next, we need to select a table. To keep it simple for now, pick the same table from the last example. Select Census_Designated_Places_2020 from the list of tables. One advantage of working in the SQL Query Builder is that we are not limited to data from a single table or schema. *Figure 6.14* is a snippet of the tables you will see when you scroll down (depending on how many tables you have created in your PostgreSQL database):

Tables

acs2020_5yr_GINI

buildingFP

buildingp

census2020__block

dec2020_GroupQuarters_P5

dohmh

nyc_census_blocks

nyc_neighborhoods

nyc_streets

nyc_subway_stations

streetNYC

streetrating

affordable_housing

Figure 6.14 – Clicking on the Tables drop-down menu to view the data

The best part of working in the SQL Query Builder is the ability to select only columns you are interested in. This will be a lifesaver for census tables and some of the more complex datasets in the final chapters. When you select **Columns** and add the WHERE clause, the SQL window will update with only the requested tables, as shown in the following screenshot:

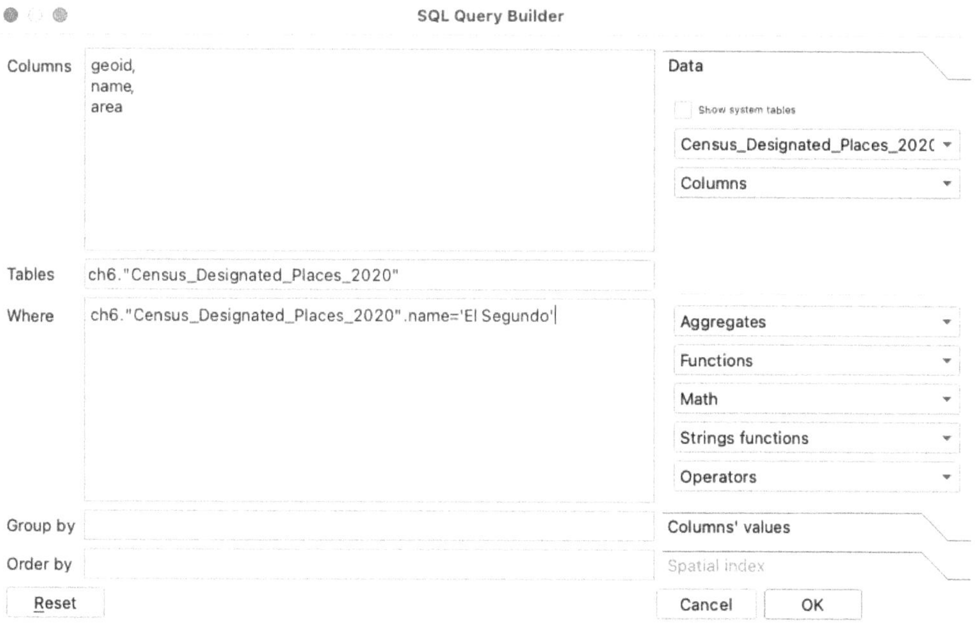

Figure 6.15 – SQL query builder in QGIS

Notice in the query results you have the three columns that you specified now displayed in *Figure 6.16*. This renders the identical output as we saw in *Figure 6.15*:

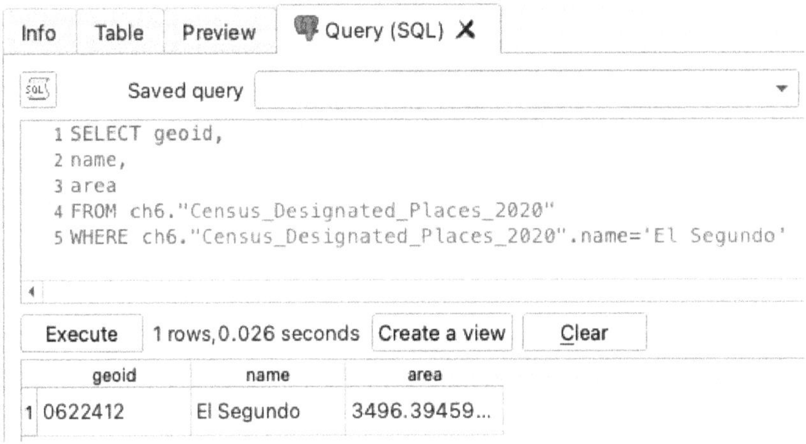

Figure 6.16 – The filtered table from the SQL query builder: QGIS

Now, let's filter by a numeric value and only see census tracts that meet the criteria we select. A simple one-variable, one-table query can be written in the query builder. You can choose the entire set of values or a sample, but to see the full range of values, I decided to scroll through and pick a random value. By testing the query, you can see that there will be enough rows returned to make it interesting.

This is the code we can write in the **DB Manager** query window:

```
SELECT ch6."Census_Designated_Places_2020".*
FROM ch6."Census_Designated_Places_2020",ch6."Below_Poverty__
census_tract_"
WHERE ST_Within(ch6."Census_Designated_Places_2020".
geom,ch6."Below_Poverty__census_tract_".geom)
```

Or we can use the query builder, as shown in *Figure 6.17*. Using the SQL query builder, you can easily filter the data to only include income below a certain value:

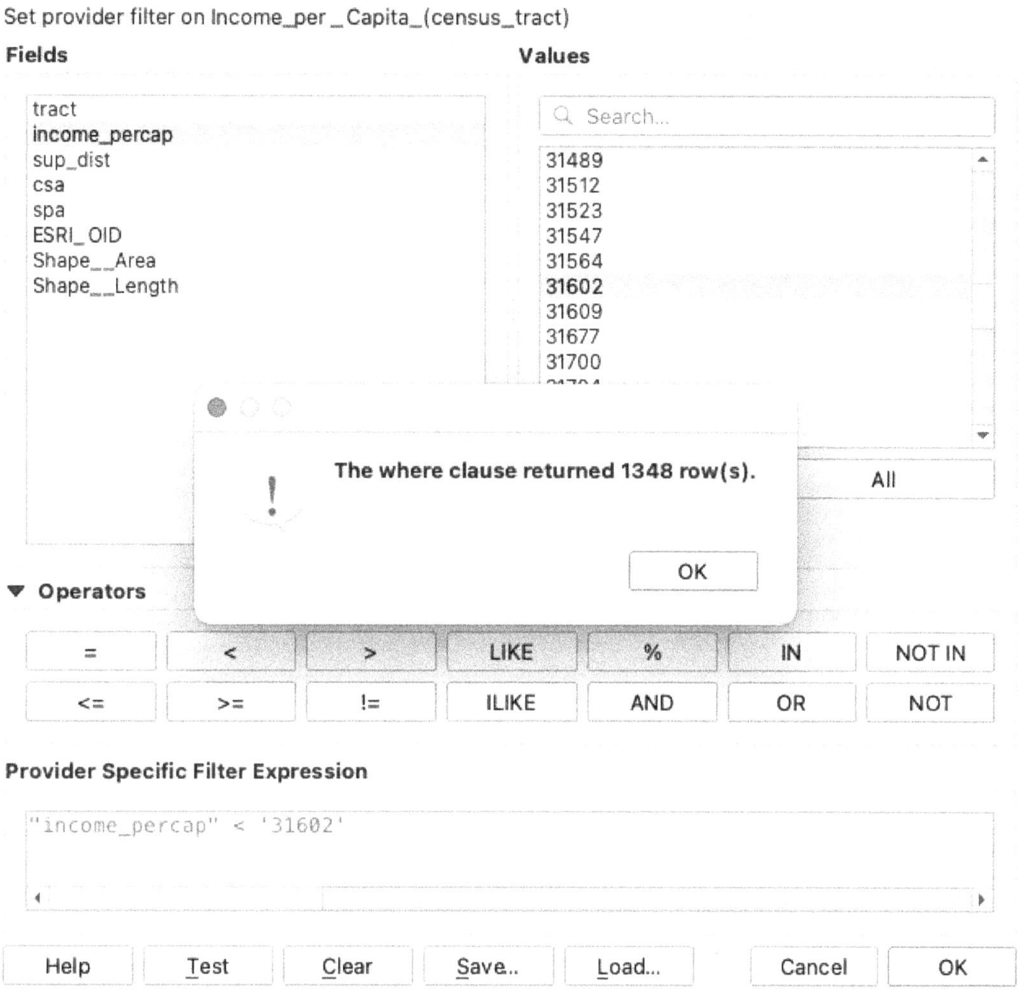

Set provider filter on Income_per _ Capita_(census_tract)

**Fields**

tract
income_percap
sup_dist
csa
spa
ESRI_ OID
Shape___Area
Shape___Length

**Values**

Q Search...

31489
31512
31523
31547
31564
**31602**
31609
31677
31700

The where clause returned 1348 row(s).

All

OK

▼ **Operators**

| = | < | > | LIKE | % | IN | NOT IN |
| <= | >= | != | ILIKE | AND | OR | NOT |

**Provider Specific Filter Expression**

```
"income_percap" < '31602'
```

| Help | Test | Clear | Save.. | Load... | Cancel | OK |

Figure 6.17 – Filtering data where census tracts have an income per capita below $31,602

Now, when we load the layer, we can update and view the census tracts with income per capita below $31,602, as shown in *Figure 6.18*:

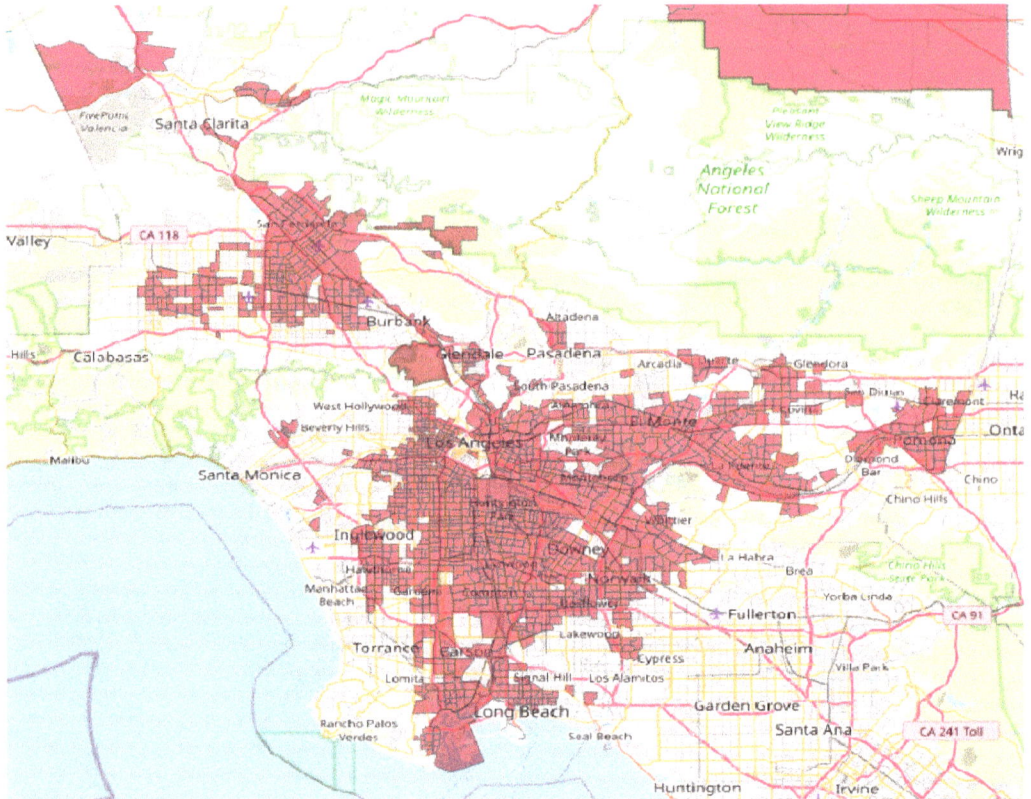

Figure 6.18 – Querying income levels by census tract in Los Angeles County

Census data is notably complicated to locate and label properly. SQL queries are well suited to making the process run more smoothly.

## Exploring census data in the SQL Query Builder

For the remaining chapter, let's explore the population dynamics in Los Angeles County—specifically, declines or increases in the Hispanic population. The folder download included a JSON file. We need to identify populations of interest from the dense JSON file you see in *Figure 6.19*:

and Alaska Native; Asian; Some other race (2010)", "P0040058_2010": "P4-58: White; American Indian and Alaska Native; Native Hawaiian and Other Pacific Islander; Some other race (2010)", "P0040059_2010": "P4-59: White; Asian; Native Hawaiian and Other Pacific Islander; Some other race (2010)", "P0040060_2010": "P4-60: Black or African American; American Indian and Alaska Native; Asian; Native Hawaiian and Other Pacific Islander (2010)", "P0040061_2010": "P4-61: Black or African American; American Indian and Alaska Native; Asian; Some other race (2010)", "P0040062_2010": "P4-62: Black or African American; American Indian and Alaska Native; Native Hawaiian and Other Pacific Islander; Some other race (2010)", "P0040063_2010": "P4-63: Black or African American; American Indian and Alaska Native; Native Hawaiian and Other Pacific Islander; Some other race (2010)", "P0040064_2010": "P4-64: American Indian and Alaska Native; Asian; Native Hawaiian and Other Pacific Islander; Some other race (2010)", "P0040065_2010": "P4-65: Population of five races (2010)", "P0040066_2010": "P4-66: White; Black or African American; American Indian and Alaska Native; Asian; Native Hawaiian and Other Pacific Islander (2010)", "P0040067_2010": "P4-67: White; Black or African American; American Indian and Alaska Native; Asian; Some other race (2010)", "P0040068_2010": "P4-68: White; Black or African American; American Indian and Alaska Native; Native Hawaiian and Other Pacific Islander; Some other race (2010)", "P0040069_2010": "P4-69: White; Black or African American; Asian; Native Hawaiian and Other Pacific Islander; Some other race (2010)", "P0040070_2010": "P4-70: White; American Indian and Alaska Native; Asian; Native Hawaiian and Other Pacific Islander; Some other race (2010)", "P0040071_2010": "P4-71: Black or African American; American Indian and Alaska Native; Asian; Native Hawaiian and Other Pacific Islander; Some other race (2010)", "P0040072_2010": "P4-72: Population of six races (2010)", "P0040073_2010": "P4-73: White; Black or African American; American Indian and Alaska Native; Asian; Native Hawaiian and Other Pacific Islander; Some other race (2010)", "P0040001_pct_chg": "P4-1: Total (% change)", "P0040002_pct_chg": "P4-2: Hispanic or Latino (% change)", "P0040003_pct_chg": "P4-3: Not Hispanic or Latino (% change)", "P0040004_pct_chg": "P4-4: Population of one race (% change)", "P0040005_pct_chg": "P4-5: White alone (% change)", "P0040006_pct_chg": "P4-6: Black or African American alone (% change)", "P0040007_pct_chg": "P4-7: American Indian and Alaska Native alone (% change)", "P0040008_pct_chg": "P4-8: Asian alone (% change)", "P0040009_pct_chg": "P4-9: Native Hawaiian and Other Pacific Islander alone (% change)", "P0040010_pct_chg": "P4-10: Some other race alone (% change)", "P0040011_pct_chg": "P4-11: Population of two or more races (% change)", "P0040012_pct_chg": "P4-12: Population of two races (% change)", "P0040013_pct_chg": "P4-13: White; Black or African American (% change)", "P0040014_pct_chg": "P4-14: White; American Indian and Alaska Native (% change)", "P0040015_pct_chg": "P4-15: White; Asian (% change)", "P0040016_pct_chg": "P4-16: White; Native Hawaiian and Other Pacific Islander (% change)", "P0040017_pct_chg": "P4-17: White; Some other race (% change)", "P0040018_pct_chg": "P4-18: Black or African American; American Indian and Alaska Native (% change)", "P0040019_pct_chg": "P4-19: Black or African American; Asian (% change)", "P0040020_pct_chg": "P4-20: Black or African American; Native Hawaiian and Other Pacific Islander (% change)", "P0040021_pct_chg": "P4-21: Black or African American; Some other race (% change)", "P0040022_pct_chg": "P4-22: American Indian and Alaska Native; Asian (% change)", "P0040023_pct_chg": "P4-23: American Indian and Alaska Native; Native Hawaiian and Other Pacific Islander (% change)", "P0040024_pct_chg": "P4-24: American Indian and Alaska Native; Some other race (% change)", "P0040025_pct_chg": "P4-25: Asian; Native Hawaiian and Other Pacific Islander (% change)", "P0040026_pct_chg": "P4-26: Asian; Some other race (% change)", "P0040027_pct_chg": "P4-27: Native Hawaiian and Other Pacific Islander; Some other race (% change)", "P0040029_pct_chg": "P4-29: White; Black or African American; American Indian and Alaska Native (% change)", "P0040030_pct_chg": "P4-30: White; Black or African American; Some other race (% change)", "P0040031_pct_chg": "P4-31: White; Black or African American; Native Hawaiian and Other Pacific Islander (% change)", "P0040032_pct_chg": "P4-32: White; Black or African American; Some other race (% change)", "P0040033_pct_chg": "P4-33: White; American Indian and Alaska Native; Asian (% change)", "P0040034_pct_chg": "P4-34: White; American Indian and Alaska Native; Native Hawaiian and Other Pacific Islander (% change)", "P0040035_pct_chg": "P4-35: White; American Indian and Alaska Native; Some other race (% change)", "P0040036_pct_chg": "P4-36: White; Asian; Native Hawaiian and Other Pacific Islander (% change)", "P0040037_pct_chg": "P4-37: White; Asian; Some other race (% change)", "P0040038_pct_chg": "P4-38: White; Native Hawaiian and Other Pacific Islander; Some other race (% change)", "P0040039_pct_chg": "P4-39: Black or African American; American Indian and Alaska Native; Asian (% change)", "P0040040_pct_chg": "P4-40: Black or African American; American Indian and Alaska Native; Native Hawaiian and Other Pacific Islander (% change)", "P0040041_pct_chg": "P4-41: Black or African American; American Indian and Alaska Native; Some other race (% change)", "P0040042_pct_chg": "P4-42: Black or African American; Asian; Native Hawaiian and Other Pacific Islander (% change)", "P0040043_pct_chg": "P4-43: Black or African American; Asian; Some other race (% change)", "P0040044_pct_chg": "P4-44: Black or African American; Native Hawaiian and Other Pacific Islander; Some other race (% change)", "P0040045_pct_chg": "P4-45: American Indian and Alaska Native; Asian; Native Hawaiian and Other Pacific Islander (% change)", "P0040046_pct_chg": "P4-46: American Indian and Alaska Native; Asian; Some other race (% change)", "P0040047_pct_chg": "P4-47: American Indian and Alaska Native; Native Hawaiian and Other Pacific Islander; Some other race (% change)", "P0040048_pct_chg": "P4-48: Asian; Native Hawaiian and Other Pacific Islander; Some other race (% change)", "P0040049_pct_chg":

Figure 6.19 – Demographic metadata JSON file Hispanic/non-Hispanic aged over 18

The area highlighted in *Figure 6.19* is the percent change in population from the 2010 census to the 2020 census. In addition, I captured the total population and total population: `Hispanic Latino for 2020, "P0040001_2020": "P4-1: Total (2020)", "P0040002_2020": "P4-2: Hispanic or Latino (2020)".`

Renaming the columns in census tables is quite a task. The codes identify data products available from the census. For example, `B` in the column heading in *Figure 6.20* is used for detailed estimates for base or detailed tables. The next two characters indicate a table topic, commuting and traveling to work (in this case, `08`), and the next three characters are unique identifiers for a topic within a topic, such as poverty status. When working on local projects, it is possible to create aliases in the attribute table window.

Downloading the ACS 2021 1-year data from *Census Reporter* (GitHub `https://github.com/PacktPublishing/Geospatial-Analysis-with-SQL` you can rename columns, as shown in *Figure 6.20*. The `metadata.json` file within the folder provides information about the table names. To keep the dataset manageable, I selected the column reporting below 100 percent poverty.

> **Note**
> This data was taken from the U.S. *Census Bureau (2021). Means of Transportation to Work by Poverty Status in the Past 12 Months, American Community Survey 1-Year Estimates.* Retrieved from `https://censusreporter.org`.

Briefly, the federal poverty level is determined each year by the **Department of Health and Human Services (DHHS)** to establish a minimum standard income amount needed to provide for a family. When a family earns less than this amount, the family becomes eligible for certain benefits to assist in meeting basic needs. Higher thresholds are provided for larger households. In the data provided by the census, you may notice higher percentages such as 150 and 200. These levels are calculated by dividing income by the poverty guideline for the year and multiplying by 100, and although these are above the minimum thresholds, they may be eligible for additional resources or subsidies.

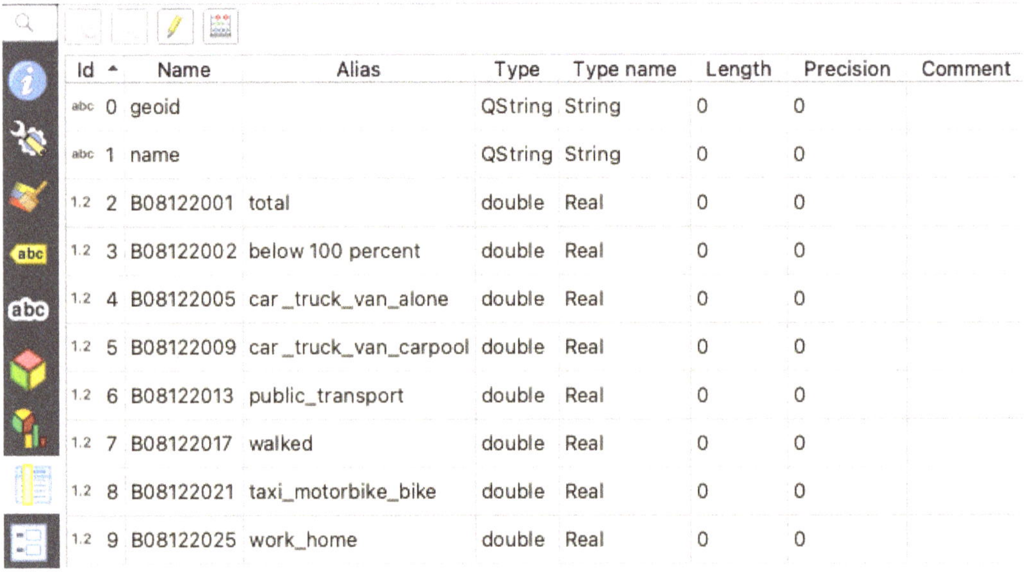

| Id ▲ | Name | Alias | Type | Type name | Length | Precision | Comment |
|---|---|---|---|---|---|---|---|
| abc 0 | geoid | | QString | String | 0 | 0 | |
| abc 1 | name | | QString | String | 0 | 0 | |
| 1.2 2 | B08122001 | total | double | Real | 0 | 0 | |
| 1.2 3 | B08122002 | below 100 percent | double | Real | 0 | 0 | |
| 1.2 4 | B08122005 | car_truck_van_alone | double | Real | 0 | 0 | |
| 1.2 5 | B08122009 | car_truck_van_carpool | double | Real | 0 | 0 | |
| 1.2 6 | B08122013 | public_transport | double | Real | 0 | 0 | |
| 1.2 7 | B08122017 | walked | double | Real | 0 | 0 | |
| 1.2 8 | B08122021 | taxi_motorbike_bike | double | Real | 0 | 0 | |
| 1.2 9 | B08122025 | work_home | double | Real | 0 | 0 | |

Figure 6.20 – Creating aliases in the Manage fields window

Scroll down to the next button in the vertical menu, and you will be able to update the alias on any field. I use this during the exploratory phase so that I can keep track of the different column headings, especially when they are coded and can be difficult to decipher.

*Figure 6.21* shows the window for adding an alias in the **General** settings. Aliases, though, will not be transferred to the source data but are also not restricted by character lengths, so can be useful within a project or project layer:

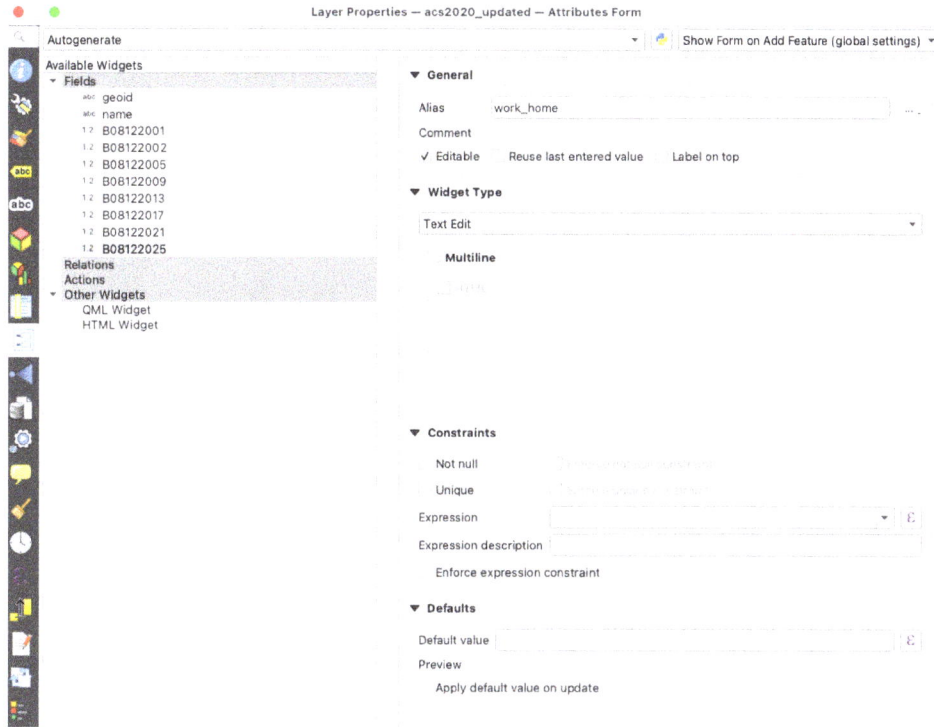

Figure 6.21 – The attributes form for adding aliases in the Layer Properties window

Now, when you access the attribute table, you have the updated columns with the aliases, as shown in *Figure 6.22*. I recommend first dropping any fields not relevant to your analyses. In the **Processing Toolbox** (**View | Panels**) setting, select **Vector Table** and **Drop field(s)**:

acs2020_updated — Features Total: 2499, Filtered: 2499, Selected: 0

| geoid | name | total | below 100 percent | ar_truck_van_alor | r_truck_van_carpc | public_transport | walked | axi_motorbike_bik | work_home |
|-------|------|-------|-------|-------|-------|-------|-------|-------|-------|
| 05000US06... | Los Angeles ... | 4761291 | 300116 | 3445866 | 452524 | 257217 | 112649 | 110808 | 382227 |
| 14000US060... | Census Tract... | 1953 | 79 | 1586 | 223 | 60 | 0 | 18 | 66 |
| 14000US060... | Census Tract... | 1997 | 27 | 1722 | 103 | 8 | 0 | 34 | 130 |
| 14000US060... | Census Tract... | 1745 | 131 | 1372 | 83 | 105 | 16 | 67 | 102 |
| 14000US060... | Census Tract... | 1631 | 58 | 1247 | 202 | 56 | 43 | 72 | 11 |
| 14000US060... | Census Tract... | 1165 | 296 | 940 | 116 | 0 | 30 | 14 | 65 |
| 14000US060... | Census Tract... | 1671 | 58 | 1315 | 120 | 29 | 0 | 0 | 207 |
| 14000US060... | Census Tract... | 1567 | 16 | 1098 | 262 | 49 | 0 | 39 | 119 |
| 14000US060... | Census Tract... | 845 | 24 | 628 | 67 | 9 | 27 | 14 | 100 |
| 14000US060... | Census Tract... | 1694 | 25 | 1177 | 118 | 43 | 16 | 2 | 338 |
| 14000US060... | Census Tract... | 1013 | 67 | 803 | 88 | 11 | 14 | 82 | 15 |
| 14000US060... | Census Tract... | 1656 | 91 | 1483 | 134 | 0 | 0 | 0 | 39 |

Figure 6.22 – Attribute table with updated field names (aliases)

I noticed that when I imported the table into the schema, the aliases were not retained. *Figure 6.23* has the source data columns. Let's explore how to make these names permanent:

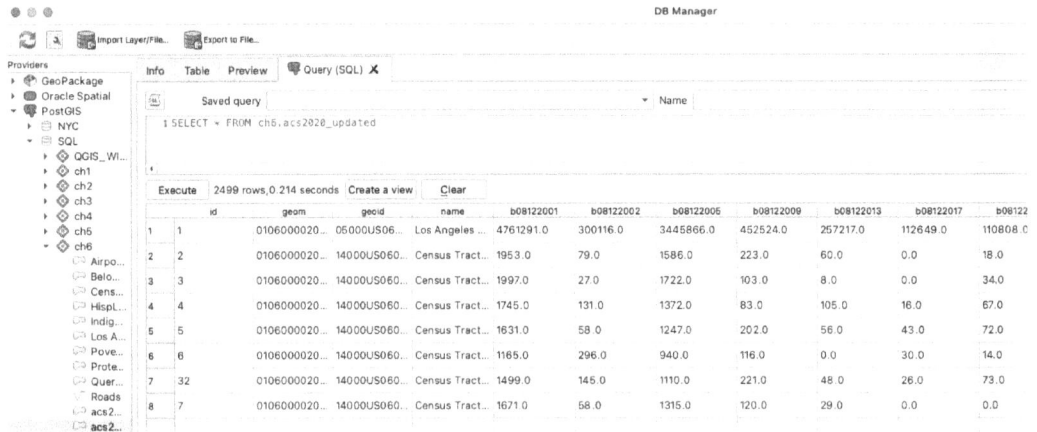

Figure 6.23 – Alias for field names not in the schema

Whenever you want to edit field names, or drop or add any field(s) to a table, return to the **Processing Toolbox** Panel and search for `Vector table`.

*Figure 6.24* displays the window for renaming fields in your table. Enter a folder name under **Renamed** if you want a permanent layer. Notice the default, [**Create temporary layer**]. In the absence of a saved layer, you will notice the icon to the right of your layers in the panel in *Figure 6.24*. The scratch layers will not be saved with your project. You will be prompted to save them if you close the project without making them permanent:

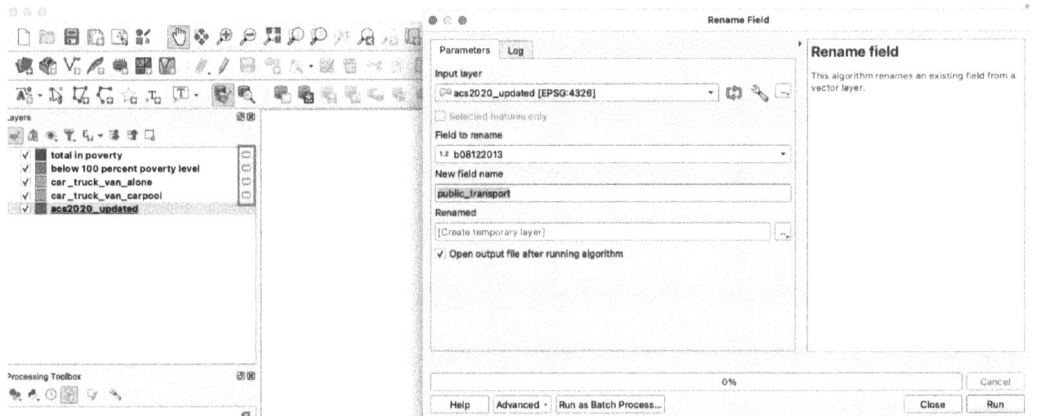

Figure 6.24 – Renaming table fields and saving them as permanent files

Let's end the chapter with a few simple queries. Run the following code to practice using an alias to simplify the code. Notice that after the FROM clause, we can use p to represent the table we are referencing. The WHERE clause is now able to use this to indicate a column in the table:

```
SELECT * FROM ch6."Below_Poverty__census_tract_"p
WHERE p.below_200fpl_pct > 48
```

We are selecting all the columns from the below-poverty census tract. This column is all of the tracts with a poverty level below the federal poverty level from the **ACS** 2021. This is 1-year data providing the *means of transportation to work by poverty status in the past 12 months*.

SQL is a powerful tool for exploring granularity in larger datasets. I notice that often when looking at demographic data, the questions are limited to population sizes, race, and income, passing over factors such as built infrastructure, environmental issues, and community barriers.

The map shown in *Figure 6.25* is showing census tracts with over 48% of households below 200 percent of the poverty threshold:

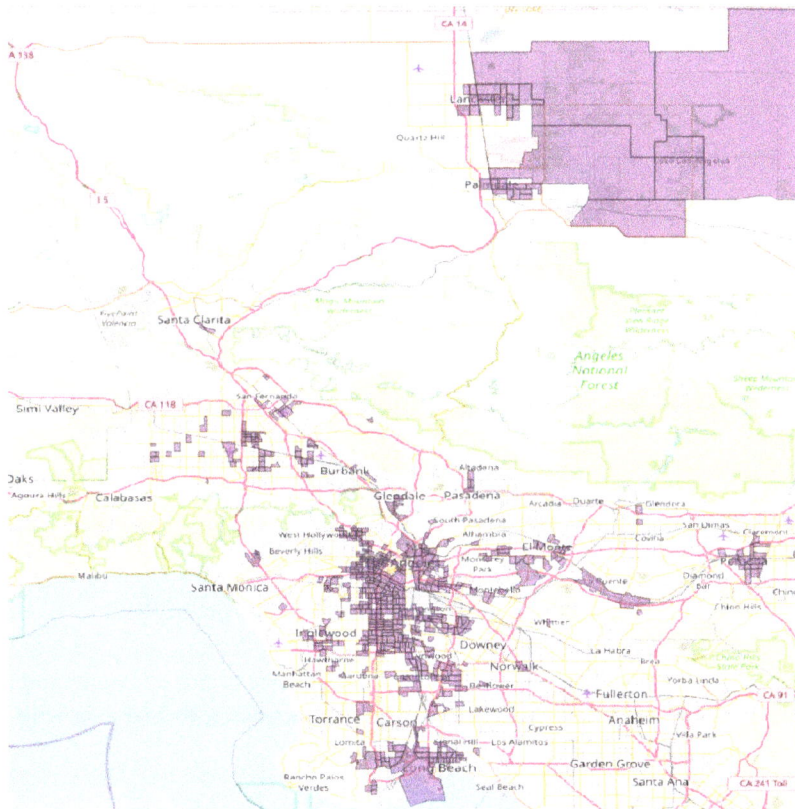

Figure 6.25 – Selecting census tracts by specific income levels

The following code also includes aliases for the Hispanic and Latino population (h) and poverty level (v). The WHERE clause can identify census tracts based on multiple calculations. We can also run this in pgAdmin to observe the data in *Figure 6.26*:

```
SELECT * FROM ch6."HispLat_LA_P2"h,ch6."Poverty_LA"v
WHERE h.p0020002_2020 >'3510' AND v.below_fpl >'450'
```

Try updating the field names in this table as well. When hovering over a census tract, information about the population can readily be observed:

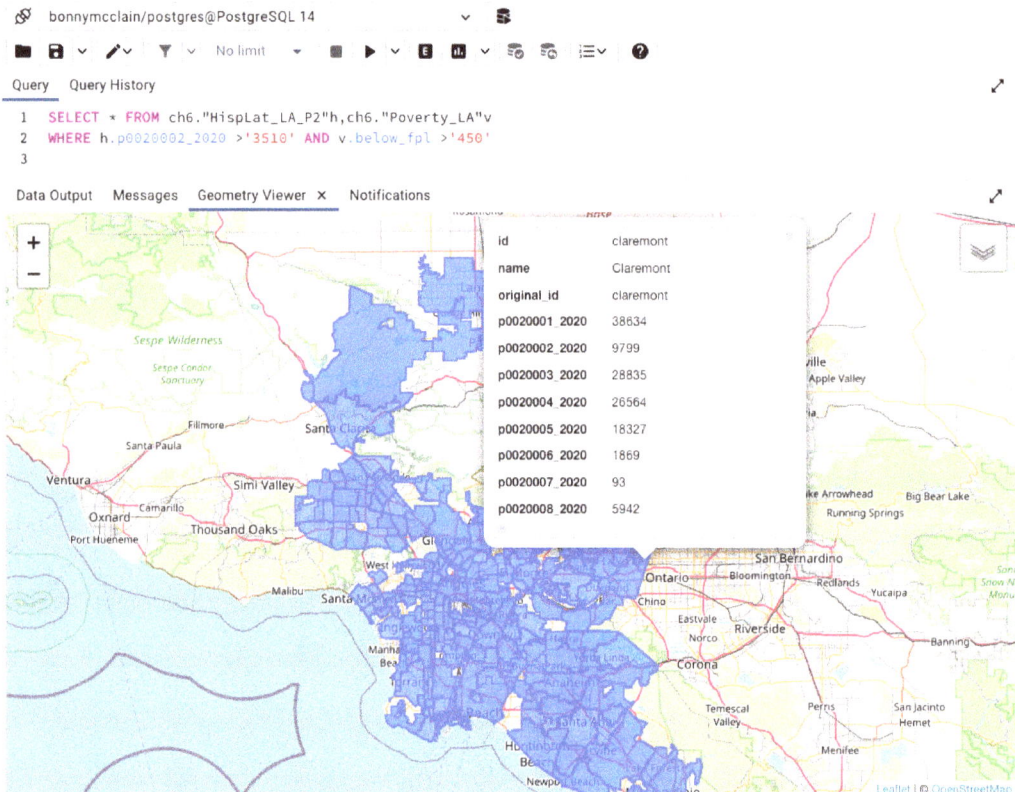

Figure 6.26 – Selecting census tracts under the federal poverty level

There are many characteristics of communities that can influence health outcomes and highlight inequities. In the final screenshot of the chapter, let's think about access to hospitals, with hospitals identified by a red plus sign inside a circle in *figure 6.27*.

What can we learn about distances traveled or time to the nearest facilities?

These are some of the complex questions we will continue to explore:

Figure 6.27 – The number of hospitals in a community can impact patient outcomes

There are different graphical query builders, and you are now familiar with two popular open source alternatives. You are ready to build your own customized workflows. For example, I tend to import datasets into QGIS and link them to pgAdmin since I work in macOS. Although I rely on the autocompletion and ease of writing SQL queries in QGIS, during the exploration of a new dataset I write and run most of my queries in pgAdmin.

There is no right way or wrong way, only the way that makes you feel most efficient and productive.

# Summary

In this chapter, we continued to learn about SQL queries as a way to filter datasets and generate focused responses to questions. The goal for writing SQL syntax is to feel conversational—building on the ability to ask specific questions. Understanding location and being able to dig a little deeper to see how "where" things are happening can often shed insight on "why" as well.

Writing conditional and aggregation data queries integrated nicely into specific workflows that you can expand as we dig deeper into the platforms that host the SQL consoles and windows.

In the next chapter, we will dig a little deeper into PostgreSQL by exploring pgAdmin in more detail and how to integrate PostGIS with QGIS.

# 7
# Exploring PostGIS for Geographic Analysis

We have been working with PostGIS to support geographic analysis (geometry and geography). Being open source and using additional spatial functions allows us to store geometric information. In this chapter, you will continue to learn how to work with the **spatial reference identifier** (**SRID**), as well as a few non-spatial functions. Information and instructions will be formatted as case studies that highlight open source data. These vignettes will correlate with how questions are formulated and encourage creativity and curiosity.

As a brief refresher, when referring to PostGIS, we mean the open source Postgres spatial extension for working with spatial databases. The main reason I rely on PostgreSQL is its natural integration with spatial information at no additional cost. This is the environment where we can execute location queries in SQL. As you will see in the next chapter, we can integrate expanded geoprocessing functions with QGIS as a sophisticated graphical interface.

In this chapter, let's dig a little deeper into PostGIS and augment your earlier introduction to creating a schema that organizes data into logical buckets. When you want to share data across projects, use the **public** schema. Your tables are within the schemas with the spatial information stored in the geom column.

We will learn about the following topics in this chapter:

- Customizing your workflows
- Importing data into QGIS Database Manager for exploration and analysis in pgAdmin
- Enabling pgAdmin to work with spatially enabled datasets
- Continuing to develop SQL queries and visualize them in the pgAdmin Geometry Viewer

# Technical requirements

The data for the exercises in this chapter can be found on GitHub at `https://github.com/PacktPublishing/Geospatial-Analysis-with-SQL`.

Please find the following datasets for working through examples:

- shapefiles_puertorico or download from website due to large file size: `http://download.geofabrik.de/north-america/us/puerto-rico.html`

- puerto-rico-190101-free (2022)

- puerto-rico-180101-free (2018)

# Configuring and authenticating PostGIS in QGIS

Although the focus of this chapter is PostGIS and the pgAdmin interface, as a macOS user, I rely on the integration of both to upload data to PostgreSQL. When initially downloading PostGIS, you are required to set a password. This is how you can link databases to QGIS and write spatial queries across platforms. Thankfully, if you do happen to forget or lose it, all is not lost.

In QGIS, head over to **Settings** and scroll down to **Options**. On the left-hand side vertical menu, go to **Authentication** and select **Configurations**, as shown in *Figure 7.1*. Here, click on **Utilities**, erase the authentication database, and recreate your configurations:

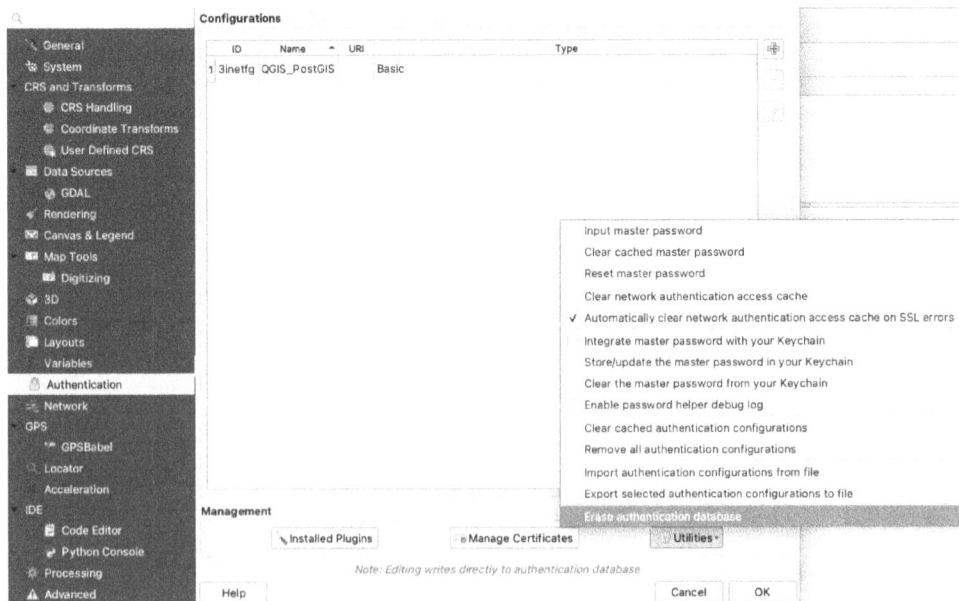

Figure 7.1 – Authentication of PostGIS configuration

Now that PostGIS has been enabled across platforms, you can work with spatially enabled datasets.

# Loading data into PostGIS

The island of Puerto Rico has been hit by island-wide power outages, storms, hurricanes, and a host of catastrophes. We can write SQL code to locate areas or locations of impact between 2019 and 2022 OSM data.

To create the dataset where we will work on structured queries with PostGIS, let's return to the *geofabrik website*, `https://download.geofabrik.de/central-america.html`. If you want to go directly to PR without the user knowing how to navigate the site and become familiar with future projects, you can go to `http://download.geofabrik.de/north-america/us/puerto-rico.html`. The dataset for Puerto Rico is located at the Central America link in the paragraph preceding the list of Central American countries. I am a big fan of **geofabrik** because not only do they provide a choice of formats for working with OSM files, but the data is updated daily.

OSM data is natively in XML format and the software you select will export the data into a format of your choosing. The shapefiles will work but require defined table structures.

*Figure 7.2*, (taken from `https://download.geofabrik.de/north-america/us/puerto-rico.html`) lists a few formats available for download, both shapefiles and `.pbf` files. I find the `.pbf` files to be easier to manage but you are free to use something else. They download in a more streamlined format and always keep geometry at the top of their minds:

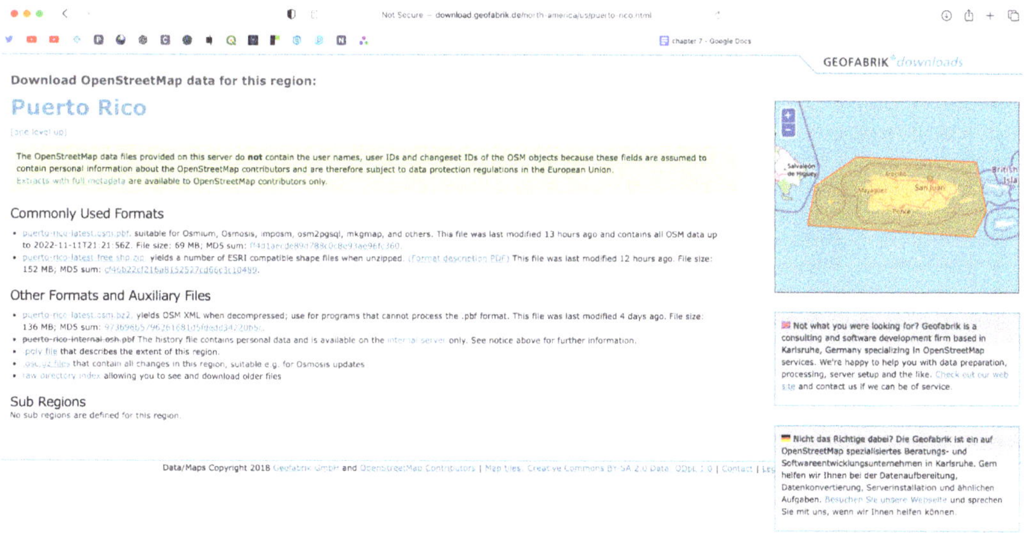

Figure 7.2 – OSM download for Puerto Rico

First, download the **protocol buffer binary format (.pbf)** due to its compatibility with OSM and its smaller size for download. Puerto Rico has a separate listing and I like this format because datasets are assigned to files by their geometry types. Features in OSM are extracted by data type within a giant database. They have been renamed by location, year, and data type.

The **Browser** panel in QGIS, as shown in *Figure 7.3*, will display the downloaded files. Notice the `osm.pbf` file and the file folders that contain shapefiles. Depending on the data you want to use, open the files; you will see the option to load them onto the canvas in QGIS. My current workflow is to simply drag the files onto the QGIS canvas. Once they are in the **Layers Panel** area, I can import them into PostgreSQL using **Database Manager**:

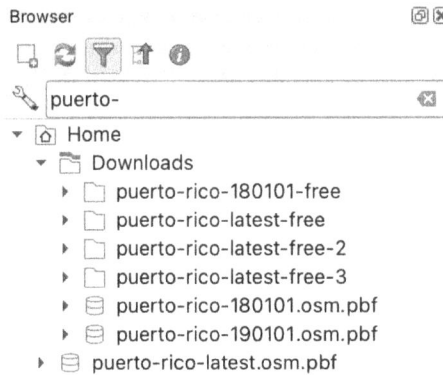

Figure 7.3 – Browser panel in QGIS to import data

By clicking on the `osm.pbf` file in *figure 7.4*, you will see that files are arranged by geometric type. I think this is cleaner and easier to organize but feel free to use the shapefile format shown in *Figure 7.5*. Because each feature is listed separately, it makes the SQL queries a bit busier, but your mileage may vary:

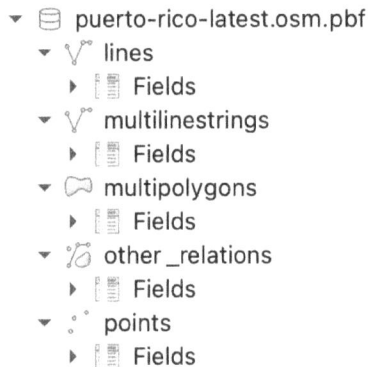

Figure 7.4 – The osm.pbf files listed by geometric type in the QGIS Browser panel

By contrast, the shape files are listed by category not geometry, as displayed in *Figure 7.5*, although `a_free` is a polygon and `free` is point data. The difference between them is whether a building is located by a point or if you can visualize the perimeter boundaries as a polygon. In OSM, point-type features are **nodes**. When you connect two or more nodes, you create a *way*, which is the line connecting the nodes. To organize multiple nodes, think about connecting nodes and ways – this represents a *multipolygon*. In OSM, these are referred to as **relations**:

puerto-rico-latest-free
  ▸ 🗺 gis_osm_landuse_a_free_1.shp
  ▸ 🗺 gis_osm_natural_a_free_1.shp
  ▸ 🗺 gis_osm_natural_free_1.shp
  ▸ 🗺 gis_osm_places_a_free_1.shp
  ▸ 🗺 gis_osm_places_free_1.shp
  ▸ 🗺 gis_osm_pofw_a_free_1.shp
  ▸ 🗺 gis_osm_pofw_free_1.shp
  ▸ 🗺 gis_osm_pois_a_free_1.shp
  ▸ 🗺 gis_osm_pois_free_1.shp
  ▸ 🗺 gis_osm_railways_free_1.shp
  ▸ 🗺 gis_osm_roads_free_1.shp
  ▸ 🗺 gis_osm_traffic_a_free_1.shp
  ▸ 🗺 gis_osm_traffic_free_1.shp
  ▸ 🗺 gis_osm_transport_a_free_1.shp
  ▸ 🗺 gis_osm_transport_free_1.shp
  ▸ 🗺 gis_osm_water_a_free_1.shp
  ▸ 🗺 gis_osm_waterways_free_1.shp

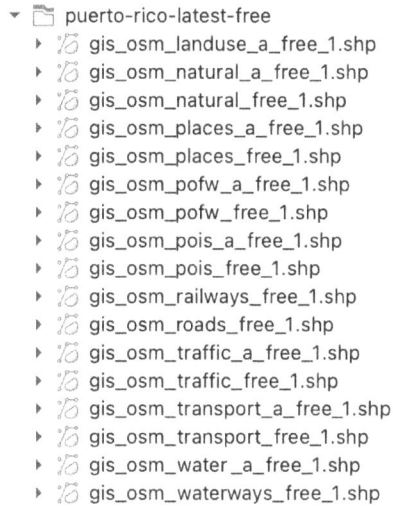

Figure 7.5 – The format of shape files within the OSM database files

The osm.pbf files also contain a **Fields** column. In *Figure 7.6*, you can see the dropdown for the column headings in the attribute table:

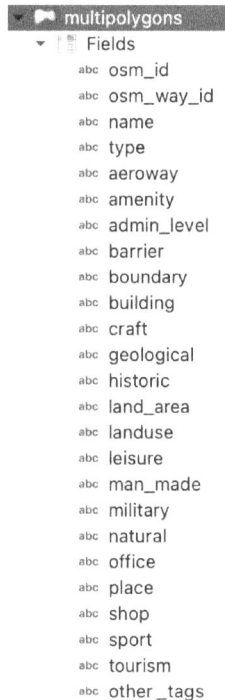

multipolygons
  ▾ Fields
      abc osm_id
      abc osm_way_id
      abc name
      abc type
      abc aeroway
      abc amenity
      abc admin_level
      abc barrier
      abc boundary
      abc building
      abc craft
      abc geological
      abc historic
      abc land_area
      abc landuse
      abc leisure
      abc man_made
      abc military
      abc natural
      abc office
      abc place
      abc shop
      abc sport
      abc tourism
      abc other_tags

Figure 7.6 – Column headings for fields in the multipolygons table

You can read more about the OSM documentation here: `https://openstreetmapjl.readthedocs.io/en/stable/index.html`.

Now that the data has been downloaded, let's connect QGIS to PostGIS so that we can explore PostGIS in pgAdmin and then also in QGIS for *Chapter 8*, *Integrating SQL with QGIS*. By scrolling down the **Browser** panel in QGIS, you can right-click and create a connection that will be visible in pgAdmin.

Now, let's look at how to get data into PostGIS.

## Importing data into PostGIS

You will need to fill in a few fields to create a connection to a database, as shown in *Figure 7.7*. The **Name** field will be the name of the connection within QGIS. Next, if you are working locally on your computer, enter localhost. The default port is 5432, and Database is the name of the database in pgAdmin. Check the boxes for listing tables with no geometry and allow QGIS to save projects in the database.

Selecting the **Test Connection** button will confirm that the connection is now valid:

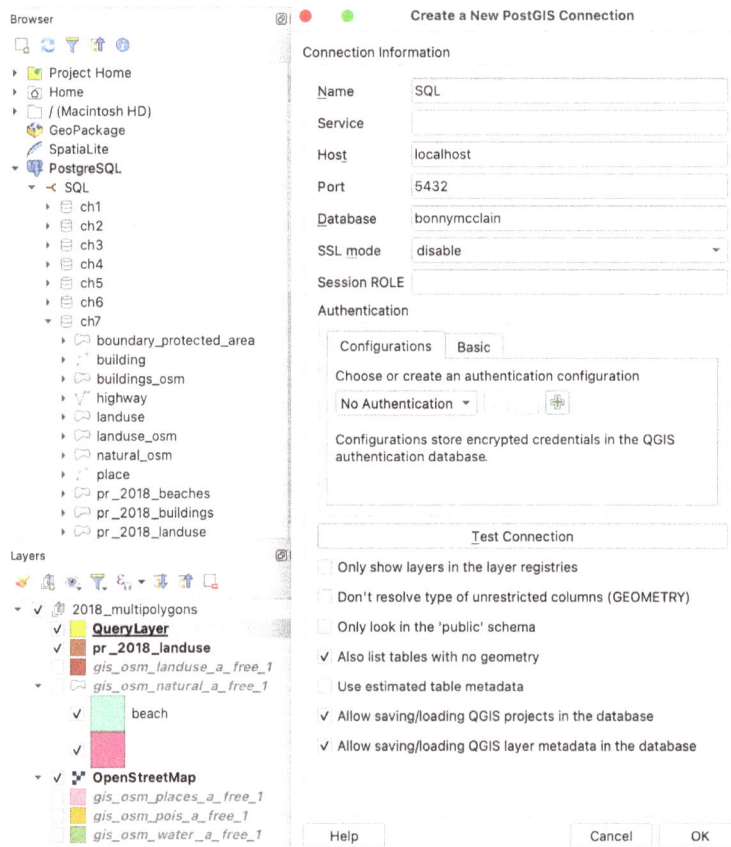

Figure 7.7 – Creating a PostGIS connection in QGIS for importing data

Feel free to rename your tables by right-clicking and entering the preferred name. Now that your data has been imported into the connection you created, which is SQL in my case, simply right-clicking, as seen in *Figure 7.8*, reveals the options for **Refresh** (to update your data) and executing SQL queries within QGIS. We will return to the query editor in *Chapter 8, Integrating SQL with QGIS*:

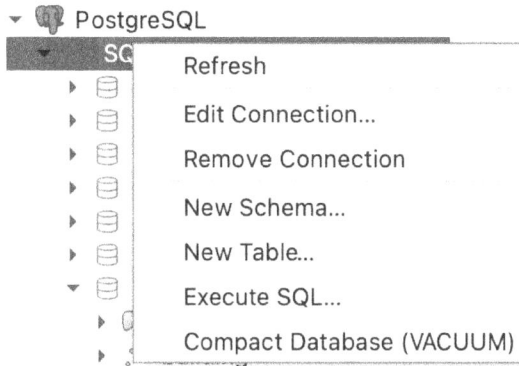

Figure 7.8 – Data layers in OSM downloaded in .pbf format

In the next section, we will return to `pgAdmin` to interact with Postgres databases and specifically PostGIS and spatial functions. Although I won't be covering the topic of database administration, it is managed within `pgAdmin` and there are additional resources available in the documentation at `https://www.pgAdmin.org/docs/pgAdmin/latest/getting_started.html`.

Now that you have accessed DB Manager to import datasets and connect to pgAdmin, you are ready to explore and analyze your data in pgAdmin.

## Discovering the PgAdmin dashboard

`pgAdmin` is a robust database administration tool with a display that you can customize according to your preferences, as shown in *Figure 7.9*. The menu bar runs across the top of the dashboard and has a series of drop-down menus to access a wide variety of **File**, **Object**, **Tools**, and **Help** commands and utilities:

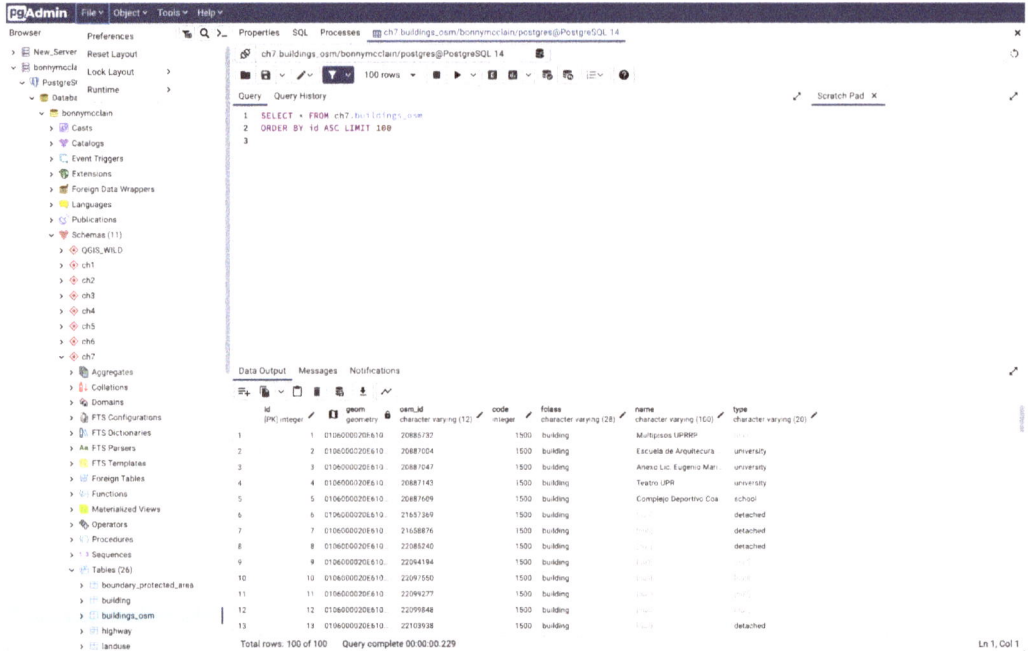

Figure 7.9 – pgAdmin dashboard display

Notice the tree control on the left vertical side of the dashboard. This is where you access your different databases, schemas, and tables. By selecting the **Preferences** dialog in the **File** menu, you can explore a few of the customizable functionalities. Remember that you will need to **CREATE** the extension for PostGIS to be able to access the spatial functions. Right-click on the database and select **Extension**. Enter PostGIS in the search box; the functions will be added to your tree structure under the database (under **Functions**).

I suggest that you explore the options across the menu bar, depending on your role and interests. *Figure 7.10* highlights the **Preferences** options and the customization options for **Graph Visualizer**. **Row Limit**, as described, is a choice between the speed of the execution of your query and the display of a graphic, chart, or map:

Preferences

- ∨ Browser
  - Display
  - Keyboard shortcuts
  - Nodes
  - Processes
  - Properties
  - Tab settings
- ∨ Dashboards
  - Display
  - Refresh rates
- ∨ Debugger
  - Keyboard shortcuts
- ∨ ERD tool
  - Keyboard shortcuts
  - Options
- ∨ Graphs
  - Display
- ∨ Miscellaneous
  - Themes
  - User language
- ∨ Paths
  - Binary paths
  - Help
- ∨ Query Tool
  - Auto completion
  - CSV/TXT Output
  - Display
  - Editor
  - Explain
  - Graph Visualiser
  - Keyboard shortcuts
  - Options
  - Results grid
  - SQL formatting
- ∨ Schema Diff
  - Display
- ∨ Storage
  - Options

Row Limit    10000

This setting specifies the maximum number of rows that will be plotted on a chart. Increasing this limit may impact performance if charts are plotted with very high numbers of rows.

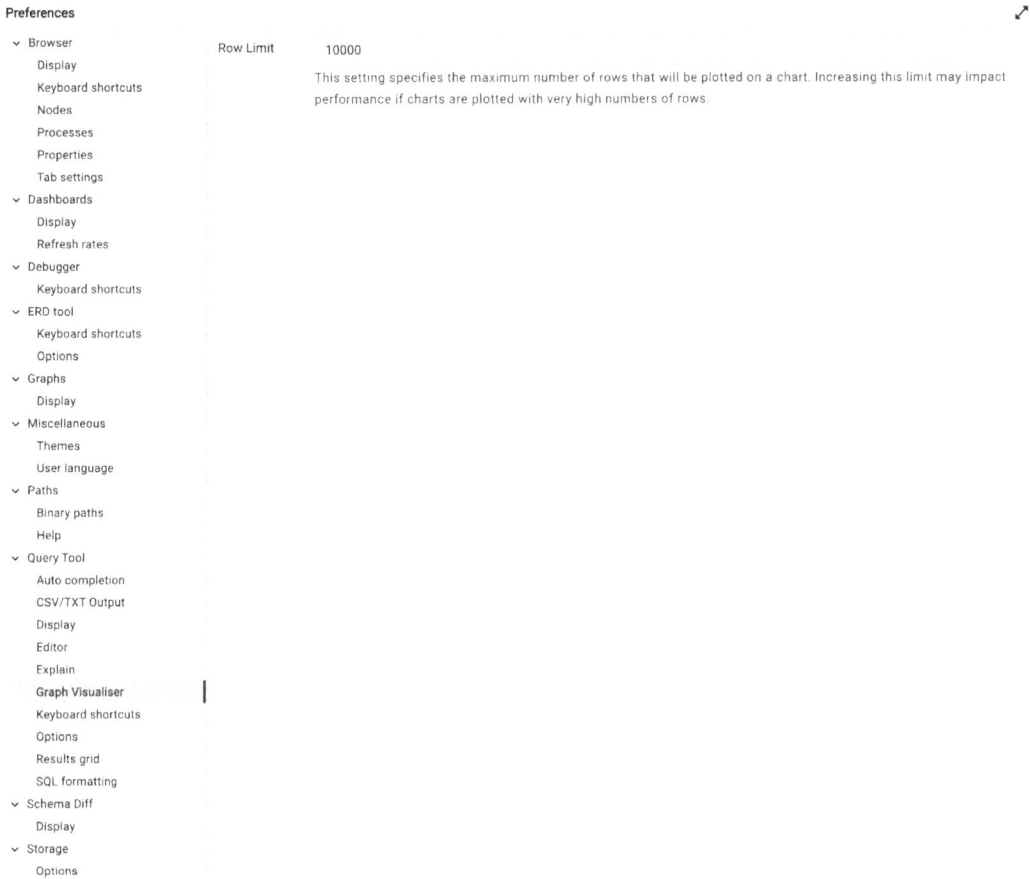

Figure 7.10 – Exploring the menu bar in pgAdmin

With that, you have imported and uploaded your data to pgAdmin. We are now ready to execute SQL queries and begin building a data story.

## Executing SQL queries in pgAdmin

Let's begin by writing a few queries that will help you become familiar with the data we downloaded and imported. I downloaded some data from the latest data distribution (2022) and also 2019 data from the archive at geofabrik.

For our query, we will select all of the columns in the table with the multipolygons by including * (see the following SQL statement). Creating an alias, mp, to refer to the table, simplifies the code for the following queries. We only want the landuse column, where the variable is equal to residential. Following major weather events, the loss of residential properties is typically profound and Puerto Rico is no exception. I invite you to also compare the differences in commercial and retail, as well as industrial. You are now able to build a richer story of the struggles in rebuilding infrastructure, many of which are unique to Puerto Rico as an island and a territory of the US. Although Puerto Ricans follow United States Federal laws, at the time of writing, they are not permitted to vote in presidential elections and do not have voting representation by Congress.

Recall that single quotes are used for variables listed in columns of a table. Because 'residential' is a variable within a column, notice the single quotes. The alias, mp, is defined in the FROM statement to simplify the query:

```
SELECT * FROM ch7.pr_2022_multipolygons mp
WHERE mp.landuse = 'residential'
```

Running the query and selecting **Geometry Viewer** displays a background map where we can view the selected results. If the background map is not displaying, you might need to transform your SRID. There are many ways to do this, but I prefer the following one:

```
ALTER TABLE table_name ALTER COLUMN geom TYPE
geometry(Point,4326) USING ST_Transform(geom,4326);
```

ST_Transform will transform coordinates in your table into the specified spatial reference system. The spatial_ref_sys table listed with your tables in pgAdmin contains a list of SRIDs.

The output of your SQL query will be like what's shown in *Figure 7.11*. **Geometry Viewer** is a convenient tool for observing the results of the query visually. From here, you can zoom into areas of interest and even select the polygons for additional information:

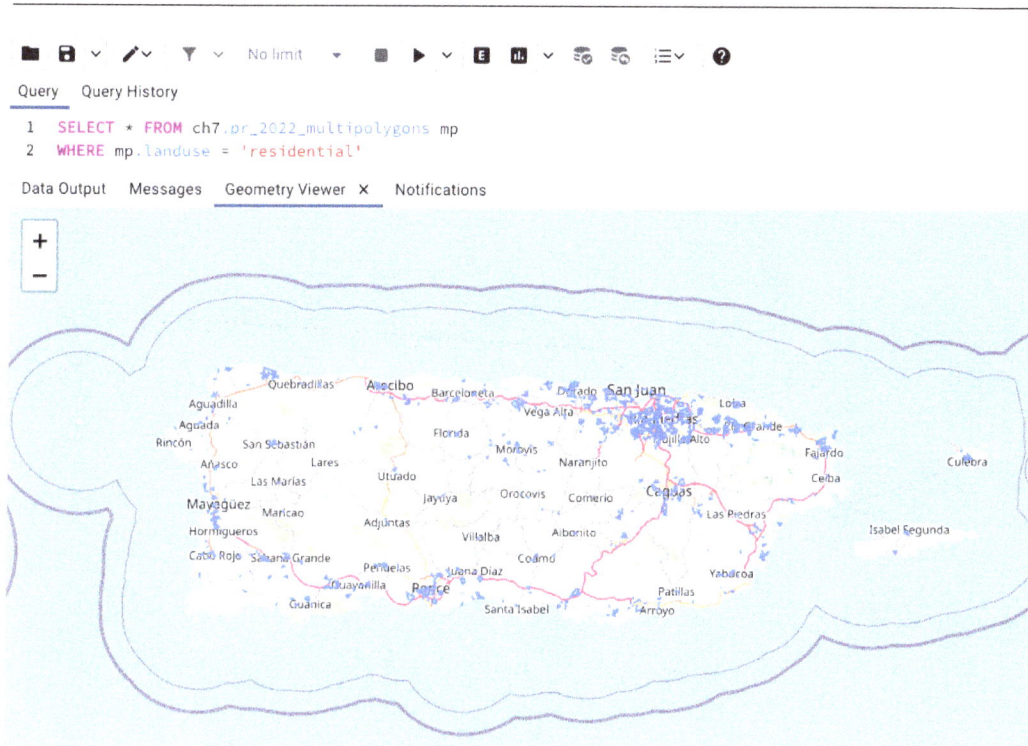

Figure 7.11 – Residential buildings in Puerto Rico in 2022

Puerto Rico has experienced several major weather events since 2017, including hurricanes Irma and Maria, as well as seismic activity. The 2022 hurricane Fiona moved through the area and the impact on infrastructure can be observed by comparing recent data with archived historical OSM data on geofabrik. You can also explore the neighboring island, Isla de Vieques, where a hospital that was lost in 2017 is still not fully rebuilt. This small island is also where the last Spanish fort in the Americas was built.

Visually, even by observing the data output, it is difficult to distinguish differences such as those in the figure provided. We will need to execute a series of SQL queries and see if we can quantify any of the results. This will be the focus of *Chapter 8, Integrating SQL with QGIS*, when we will explore raster functions.

*Figure 7.12* displays 2018 data from geofabrik that was downloaded using the available shapefiles:

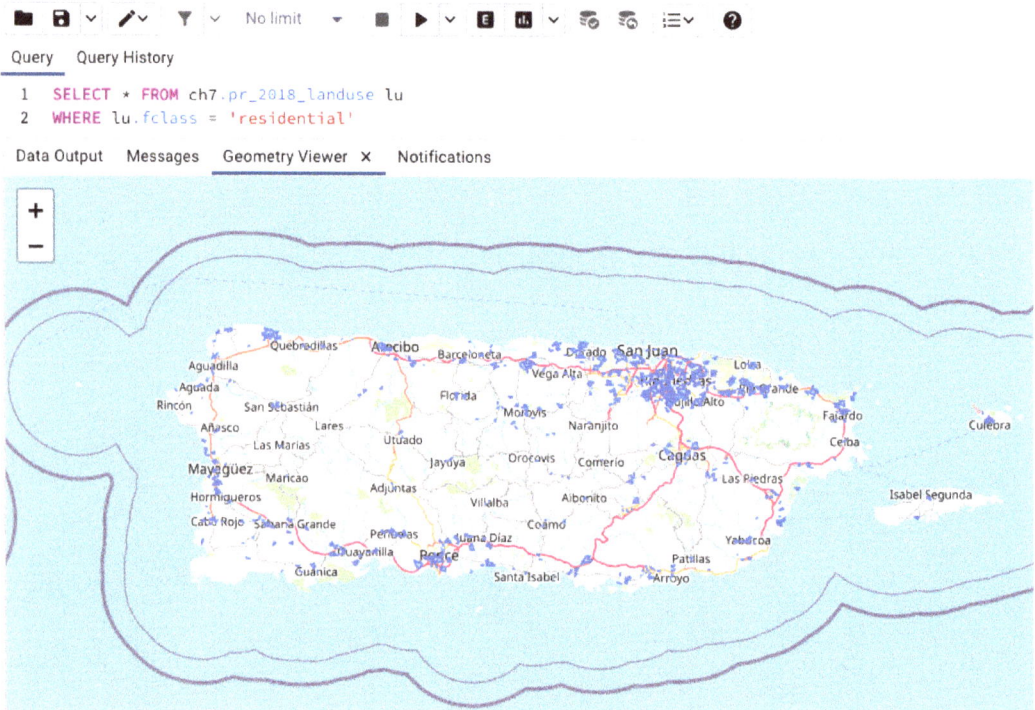

Figure 7.12 – Residential buildings in Puerto Rico in 2018

The column headings are slightly different but the overall structure of the query remains the same:

```
SELECT * FROM ch7.pr_2018_landuse lu
WHERE lu.fclass = 'residential'
```

To highlight and work with SQL syntax, it is important to identify datasets. The Puerto Rico dataset, although only a snapshot of potential queries, has a wide variety of data layers, including boundaries or polygons such as buildings or areas of interest and amenities that include facilities for use by residents or visitors. Point data includes places and lines and multilines that include routes, waterways, and highways.

# Importing additional data – power plants

For example, if we want to look at the location of `Power_Plants` in Puerto Rico, we can upload the data and select the data from Puerto Rico, as seen in *Figure 7.13*:

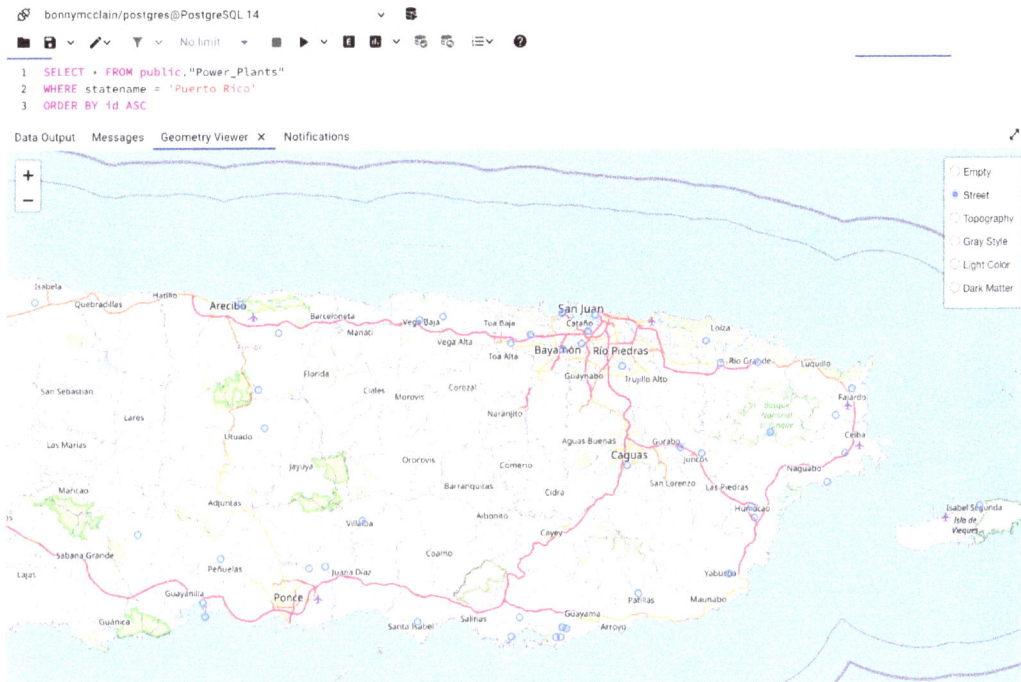

Figure 7.13 – Public power plants in Puerto Rico

Click on the layers icon at the top right-hand corner of the canvas and explore the options for basemaps that are available in `pgAdmin`:

```
SELECT * FROM public."Power_Plants"
WHERE statename = 'Puerto Rico'
ORDER BY id ASC
```

There are also times when the basemap should fade into the background to highlight another feature, such as when we look at the change in waterways after weather events, as shown in *Figure 7.14* (2019) and *Figure 7.15* (2022):

```
SELECT * FROM public.pr_multipolygons_2019
WHERE "natural" = 'water'
ORDER BY id ASC LIMIT 100000
```

··· ·

```
1   SELECT * FROM public.pr_multipolygons_2019
2   WHERE "natural" = 'water'
3   ORDER BY id ASC LIMIT 100000
```

Data Output    Messages    Geometry Viewer ✕    Notifications

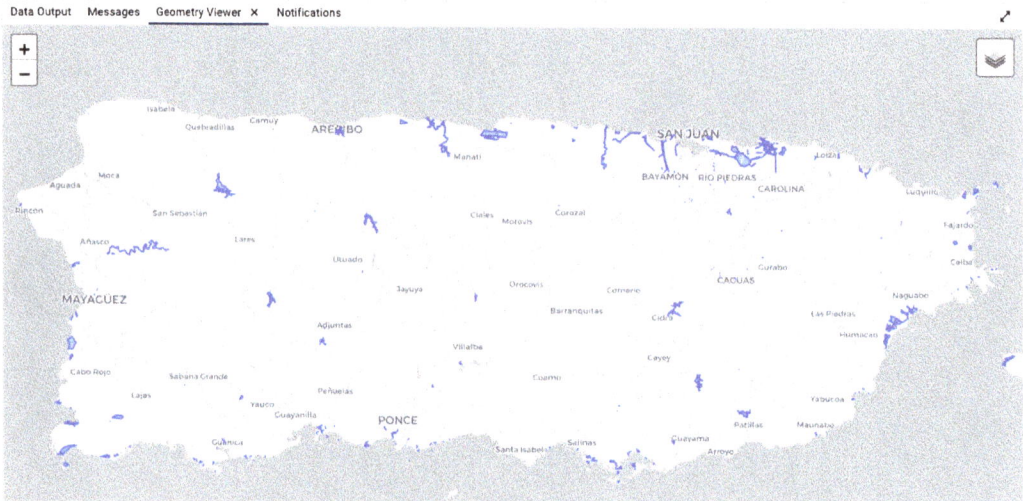

Figure 7.14 – A 2019 representation of a natural multipolygon, n=362

There are many features to explore in the datasets. Often, it is a simple query that leads to interesting insights. Notice the number of rows generated for a simple idea of quantifying insights:

```
SELECT * FROM public.pr_multipolygons_2022
WHERE "natural" = 'water'
ORDER BY id ASC LIMIT 100000
```

······

```
bonnymcclain/postgres@PostgreSQL 14
1  SELECT * FROM public.pr_multipolygons_2022
2  WHERE "natural" = 'water'
3  ORDER BY id ASC LIMIT 100000
```

Data Output   Messages   Geometry Viewer   X   Notifications

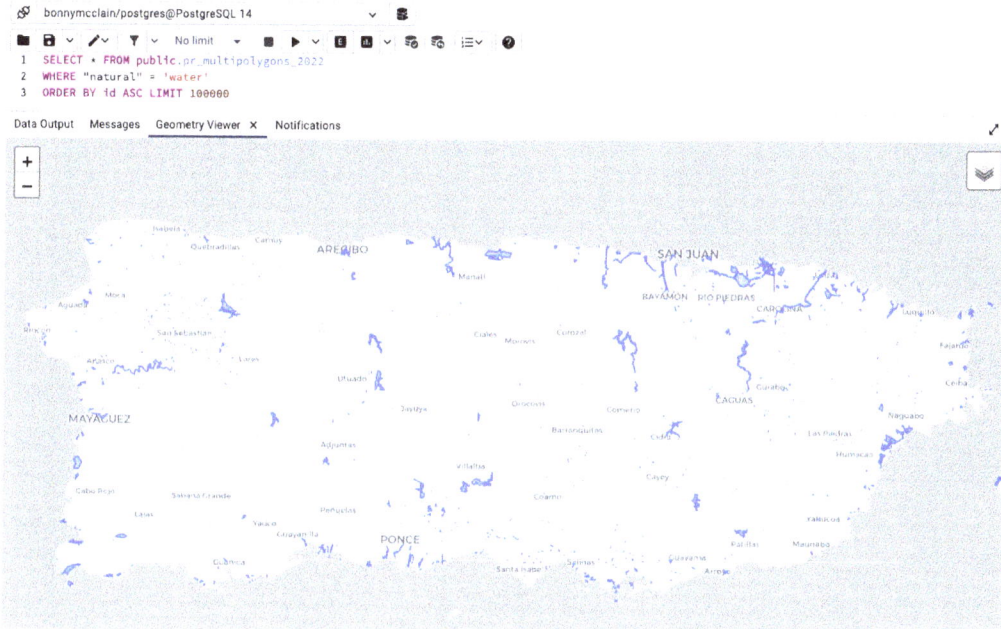

Figure 7.15 – A 2022 representation of a natural multipolygon, n=1665

This book isn't able to provide inexhaustible instructions on any technical tool or platform. My goal here is to provide a template for deeper exploration both in writing SQL queries and geospatial analysis.

You should bring a variety of data layers into any data question. This includes areas, boundaries, and building polygons combined with point data that includes named places, as well as lines and multilines for demonstrating roads and routes through and around a geographical area.

# Additional data queries – IS NOT NULL

When working with large datasets, especially OSM files, it makes sense that not every column will have an associated variable. The number of null values can be quite expansive and perhaps you don't want to include them in your analysis or visualization. Exercising caution when removing missing data is critical but in our case, for example, perhaps we only want data that includes a name for the feature we are exploring.

SQL uses an IS NOT NULL condition that simply returns TRUE when it encounters a non-NULL value; otherwise, it will return FALSE. This is applicable to SELECT, INSERT, UPDATE, or DELETE statements.

In *Figure 7.16*, we are querying `land_area`, which returns the name of the `land_area` specified. We can see that three values or polygons are returned. Let's take a look at the **Geometry Viewer** area in *Figure 7.17*:

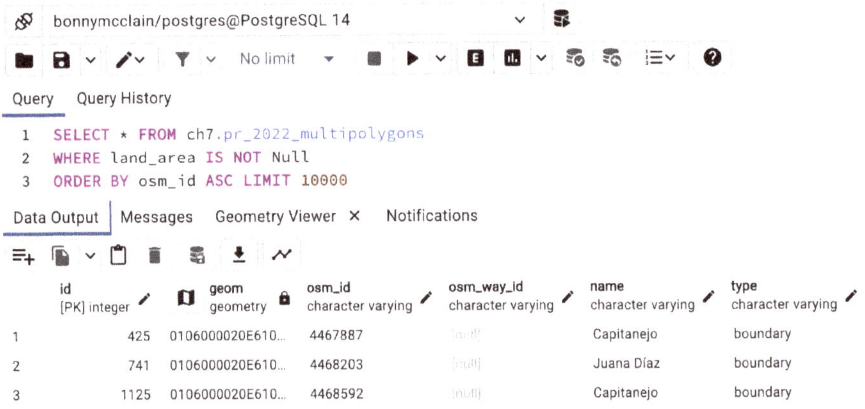

Figure 7.16 – IS NOT Null used in multipolygon dataset

When we visualize our data in the **Geometry Viewer** area, we will see the three municipalities represented. Curious as to why we only have these areas visualized, we can return to the data output and observe the rows with missing data in the **name** column.

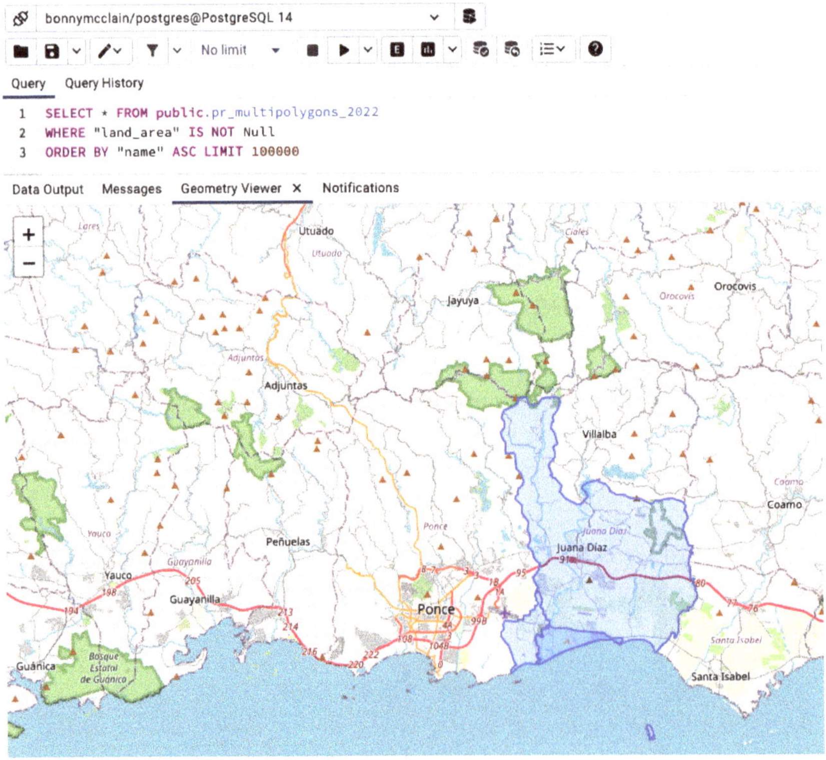

Figure 7.17 – Geometry Viewer of the IS NOT Null statement for land use in OSM

The data shows that not all of the municipalities have a name entered. Let's rerun the query and see what the output looks like if we include null values. *Figure 7.18* returns the `land_area` in the puerto rico multipolygon shapefile dataset we anticipated:

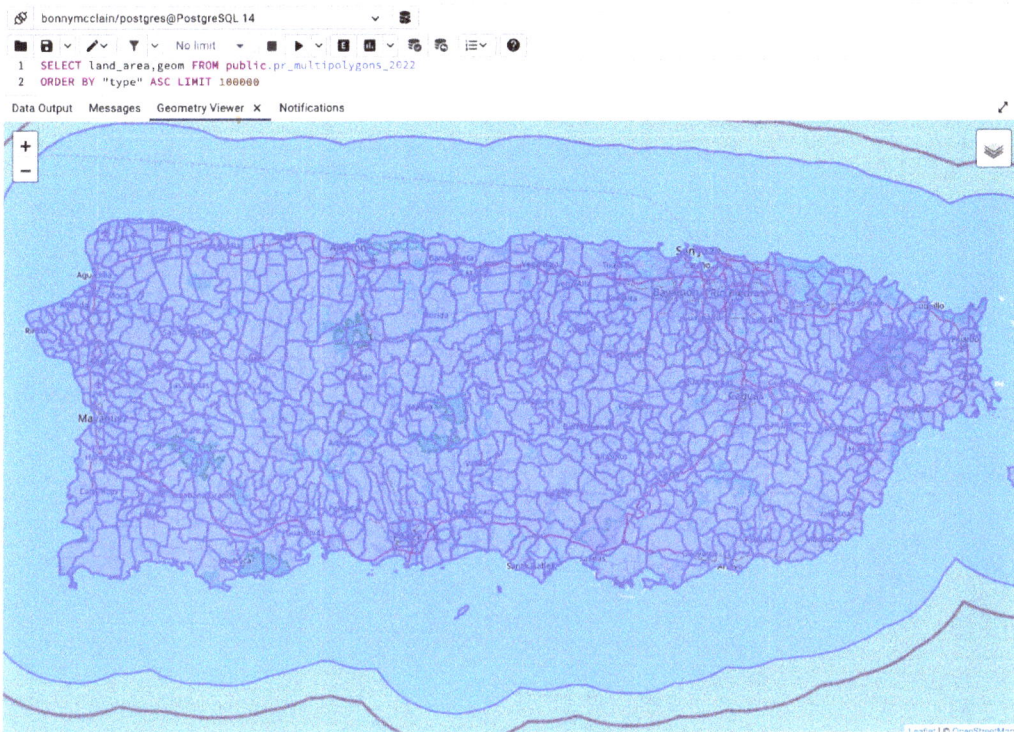

Figure 7.18 – Land_area without IS NOT Null in OSM

Explore this using `IS NOT NULL` and compare the 2022 data with the 2019 data:

```
SELECT * FROM public.pr_multilinestrings_2022
ORDER BY "type" ASC LIMIT 100000
```

The multiline data shows routes through a geographical area, so I imagined they would slowly increase in number after a natural disaster. It might be revealing to explore the rate of rebuilding and if certain municipalities seem to have priority.

What do you think the 2022 data will reveal? Simply change the table year and see what you notice. Does this change if you remove `IS NOT NULL`? Why or why not?

Upon exploring IS NOT NULL one more time in *Figure 7.19* and comparing it to the output shown in *Figure 7.20* where we remove the expression, there are a few more routes visible:

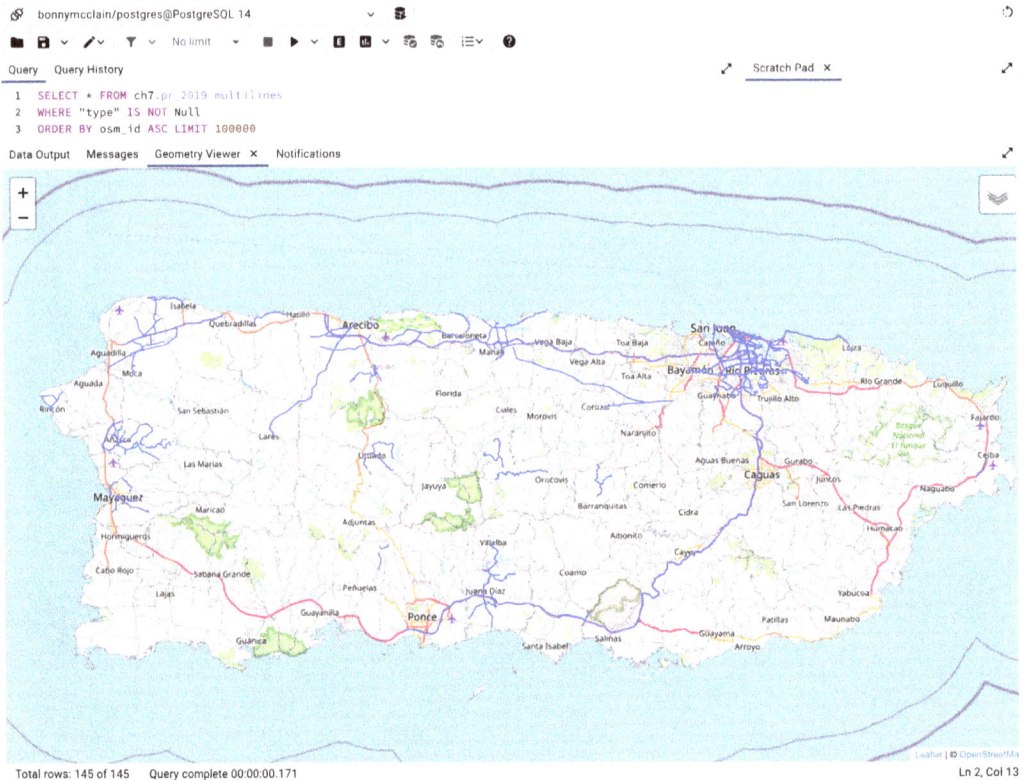

Figure 7.19 – Multilines showing routes in Puerto Rico in 2019

Here is the output comparison with *Figure 7.20*:

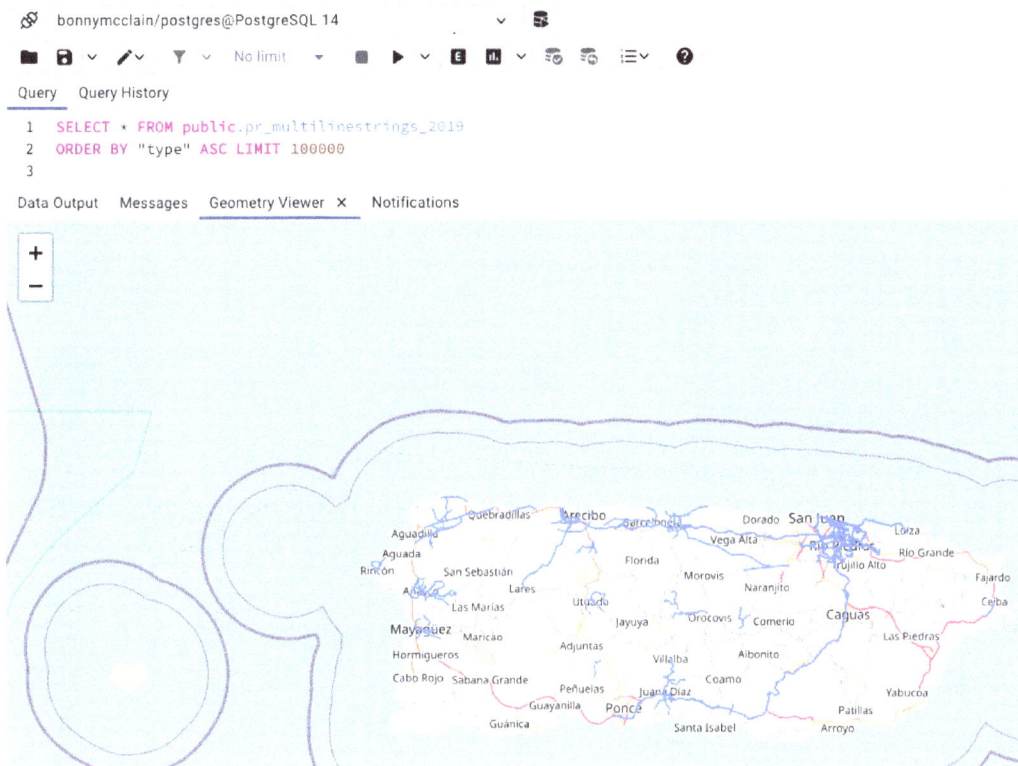

Figure 7.20 – Multilines including null values for 2019

Because OSM is a living document, one of the disadvantages includes potentially incomplete datasets. Always do a little research to also examine what you should expect to see in a dataset. For example, knowing the number of municipalities in a region or background knowledge about a transportation network or drivers of change in deforestation or military presence will help inform your analysis.

# Differentiating polygons in OSM data

A reminder of the syntax used in SQL query language helps to clear our minds as we generate complex questions within and between datasets:

- `SELECT name_attribute`: What is the attribute you are selecting from your dataset?
- `FROM name_table`: What table are you selecting this attribute from?
- `WHERE`: What conditions need to be met?
- `GROUP BY name_attribute`: Should they be sorted by name, area, or ID for example?
- `HAVING`: Additional criteria to be met, often a threshold value.

Try to recognize this template in queries you observe or create. There should be a story embedded in each query.

## Spatial queries

How does PostGIS extract information from a database? Recall that we are querying coordinates, reference systems, and other dimensions from a wide variety of datasets of different geometries that include the length of a line, the area of a polygon, or even a point location. These are collected based on specific properties of geometries.

OpenStreetMap data is often stored in a Postgres database with PostGIS extensions. This type of database provides fast access to the data and can be readily imported into QGIS to import the raw OSM data into a Postgres database.

When uploading, you will notice the **Overpass API** running in the background to extract the data from the main `OpenStreetMap` database. The API is faster than simply accessing the main OSM database server. Using SQL query language allows you to customize the subset of the data you are interested in exploring.

That brief reminder will be helpful when we approach multi-table queries again.

For example, consider the following:

```
SELECT expressions FROM table 1 JOIN table 2 ON predicate
```

I was taught this in a course I took years ago. Think of a predicate as determining if the point of your initial query is true or false. The point of joining a table on a true or false predicate is to assert something about the statement.

This predicate is based on field values replaced by spatial relationships between features in your dataset. Now, I am saying that if the expressions I select are true, I would like to join this table to a second table based on another set of assertions. You can keep adding joins like so, so long as you are considering a statement that asserts something:

```
SELECT expressions FROM table 1 JOIN table 2 ON predicate JOIN
table 3 ON predicate
......
```

Let's move on to exploring `ST_Contains` and multi-table joins.

## Exploring ST_Contains and multi-table joins

This is easier if we take an example from our data. The PostGIS extension has a host of functions for us. Return to your pgAdmin display and scroll down within one of your spatially enabled databases. You won't have functions to display until you CREATE the PostGIS extension, as we did earlier. After my last count, I had over 700 in the latest database.

Let's use ST_Contains to add a JOIN and explore a multi-table spatial query. ST_Contains accepts two geometries (a and b). If geometry b or the second geometry is located inside geometry a or the first geometry, then you have a true value. If the second geometry is not located in the first, it will return false and you won't have a value returned. Because this function returns a binary or true/false value, it can be used for a JOIN. Using these functions in this way is referred to as a **spatial join**.

We are using aliases for the multipolygon table, mp, and the points table, c. You need to select the geometries if you want to be able to visualize the output in the **Geometry Viewer** area.

The question I am formulating is about determining how many hospitals are in the cities located in the place column in the pr_points_2019 dataset. The count() function only counts non-null values but I am interested in returning all of the rows. This is the count(*) function in the query.

So, the story is to select hospitals inside the cities and tally them up! But only do so if the building is inside the city and also a hospital. Have a look at the following query:

```
SELECT mp.geom,mp.building,
count(*) as count
FROM pr_multipolygons_2019 mp
JOIN pr_points_2019 c
ON ST_Contains(mp.geom, c.geom)
WHERE building = 'hospital'
GROUP BY
mp.building,mp.geom,c.name;
```

The data output is shown in *Figure 7.21*:

Properties   SQL   Processes   🗄 bonnymcclain/...   ⊞ public.pr_point...   🗄 bonnymcclain/pos

🔗   bonnymcclain/postgres@PostgreSQL 14    ⌄   🗄

```sql
1   SELECT mp.geom,mp.building,
2   count(*) as count
3   FROM pr_multipolygons_2019 mp
4   JOIN pr_points_2019 c
5   ON ST_Contains(mp.geom, c.geom)
6   WHERE building = 'hospital'
7   GROUP BY
8   mp.building,mp.geom;
9
```

Data Output  |  Messages   Geometry Viewer ✕   Notifications

| | geom<br>geometry | building<br>character varying | count<br>bigint |
|---|---|---|---|
| 1 | 0106000020E610... | hospital | 1 |
| 2 | 0106000020E610... | hospital | 1 |
| 3 | 0106000020E610... | hospital | 2 |
| 4 | 0106000020E610... | hospital | 1 |
| 5 | 0106000020E610... | hospital | 2 |
| 6 | 0106000020E610... | hospital | 1 |
| 7 | 0106000020E610... | hospital | 1 |
| 8 | 0106000020E610... | hospital | 2 |
| 9 | 0106000020E610... | hospital | 1 |
| 10 | 0106000020E610... | hospital | 1 |
| 11 | 0106000020E610... | hospital | 1 |
| 12 | 0106000020E610... | hospital | 1 |

Figure 7.21 – pgAdmin display showing data output of n=15 hospitals

Now, we have a count of the hospitals in 2019. Our dataset reports 15 hospitals. Now, let's add the name of the hospital to the query and rerun it with our 2022 dataset, as shown in *Figure 7.22*:

Properties    SQL    Processes    🐘 bonnymcclain/...    ⊞ public.pr_point...

🔗    bonnymcclain/postgres@PostgreSQL 14                    ⌄    🔧

📁    💾 ⌄    ✏ ⌄    ▼ ⌄    No limit    ⌄    ■    ▶ ⌄    E    📊 ⌄    🔄    ⋮

Query    Query History

```sql
1  SELECT mp.geom,mp.building,c.name,
2  count(*) as count
3  FROM pr_multipolygons_2022 mp
4  JOIN pr_points_2022 c
5  ON ST_Contains(mp.geom, c.geom)
6  WHERE building = 'hospital'
7  GROUP BY
8  mp.building,mp.geom,c.name;
9
```

Data Output    Messages    Geometry Viewer ✕    Notifications

≡+    📋 ⌄    📋    🗑    🗄    ⬇    〰

|  | geom 🔒 geometry | building 🔒 character varying | name 🔒 character varying | count 🔒 bigint |
|----|----|----|----|----|
| 1 | 0106000020E610... | hospital | Ryder Memorial H... | 1 |
| 2 | 0106000020E610... | hospital | HIMA·San Pablo ... | 1 |
| 3 | 0106000020E610... | hospital | Presbyterian Com... | 1 |
| 4 | 0106000020E610... | hospital | Escuela de Artes ... | 1 |
| 5 | 0106000020E610... | hospital | Hato Rey Heliport | 1 |
| 6 | 0106000020E610... | hospital | Centro Cardiovasc... | 1 |
| 7 | 0106000020E610... | hospital | Hotel Howard Joh... | 1 |
| 8 | 0106000020E610... | hospital | Health South Reha... | 1 |
| 9 | 0106000020E610... | hospital | Administracion De... | 1 |
| 10 | 0106000020E610... | hospital | Puerto Heliport | 1 |
| 11 | 0106000020E610... | hospital | Department of Vet... | 1 |
| 12 | 0106000020E610... | hospital | Bayamon Health C... | 1 |
| 13 | 0106000020E610... | hospital | Hospital Hermano... | 1 |
| 14 | 0106000020E610... | hospital | Puerto Rico Childr... | 1 |
| 15 | 0106000020E610... | hospital | Centro Médico Wil... | 1 |
| 16 | 0106000020E610... | hospital | Healthsouth Reha... | 1 |
| 17 | 0106000020E610... | hospital | The Renal Center ... | 1 |
| 18 | 0106000020E610... | hospital | Ramon Ruiz Arnau... | 1 |
| 19 | 0106000020E610... | hospital | Laboratorios Borin... | 1 |
| 20 | 0106000020E610... | hospital | Legends Barrabrava | 1 |
| 21 | 0106000020E610... | hospital | Navitas Medical B... | 1 |

Figure 7.22 – Hospital data from 2022 multi-table join in pgAdmin, n = 32

Remember that the data is only as complete as the updating and uploading to OSM, but we can see that there are more hospitals back online in 2022, and overall, that is a good thing. We can rerun our 2019 data and add the names, remembering that our count (*) will also return null values, as shown in *Figure 7.23*:

Properties   SQL   Processes   🗄 bonnymcclain/...   ▦ public.pr_point...   🗄 bonnymcclai

🔗   bonnymcclain/postgres@PostgreSQL 14                    ⌄   🗄

📁   💾 ⌄   ✏️ ⌄   🔽 ⌄   No limit   ▾   ◼ ▶ ⌄   E   ◫ ⌄   🔄 🔄   ☰⌄

Query    Query History

```
1   SELECT mp.geom,mp.building,c.name,
2   count(*) as count
3   FROM pr_multipolygons_2019 mp
4   JOIN pr_points_2019 c
5   ON ST_Contains(mp.geom, c.geom)
6   WHERE building = 'hospital'
7   GROUP BY
8   mp.building,mp.geom,c.name;
9
```

Data Output    Messages    Geometry Viewer ✕    Notifications

=+   📋 ⌄   📋   🗑️   🗄   ⬇️   〜

|   | geom<br>geometry | building<br>character varying | name<br>character varying | count<br>bigint |
|---|---|---|---|---|
| 1 | 0106000020E610... | hospital | Presbyterian Com... | 1 |
| 2 | 0106000020E610... | hospital | Escuela de Artes ... | 1 |
| 3 | 0106000020E610... | hospital | Centro Cardiovasc... | 1 |
| 4 | 0106000020E610... | hospital | Hotel Howard Joh... | 1 |
| 5 | 0106000020E610... | hospital | Health South Reha... | 1 |
| 6 | 0106000020E610... | hospital | Administracion De... | 1 |
| 7 | 0106000020E610... | hospital | [null] | 1 |
| 8 | 0106000020E610... | hospital | Department of Vet... | 1 |
| 9 | 0106000020E610... | hospital | Bayamon Health C... | 1 |
| 10 | 0106000020E610... | hospital | Hospital Hermano... | 1 |
| 11 | 0106000020E610... | hospital | Puerto Rico Childr... | 1 |
| 12 | 0106000020E610... | hospital | Ramon Ruiz Arnau... | 1 |
| 13 | 0106000020E610... | hospital | Centro Medico Del... | 1 |
| 14 | 0106000020E610... | hospital | Hospital San Carlo... | 1 |
| 15 | 0106000020E610... | hospital | Clinica Santa Rosa | 1 |

Figure 7.23 – Identifying the names of hospitals located in the cities included in the dataset

In *Figure 7.24*, we can zoom into one of the hospitals. It can be challenging to locate them on the maps since there are so few of them and they are scattered throughout the island. The majority are located near San Juan and the larger cities. Again, when querying our datasets, we need to think about the limitations of the data we are analyzing:

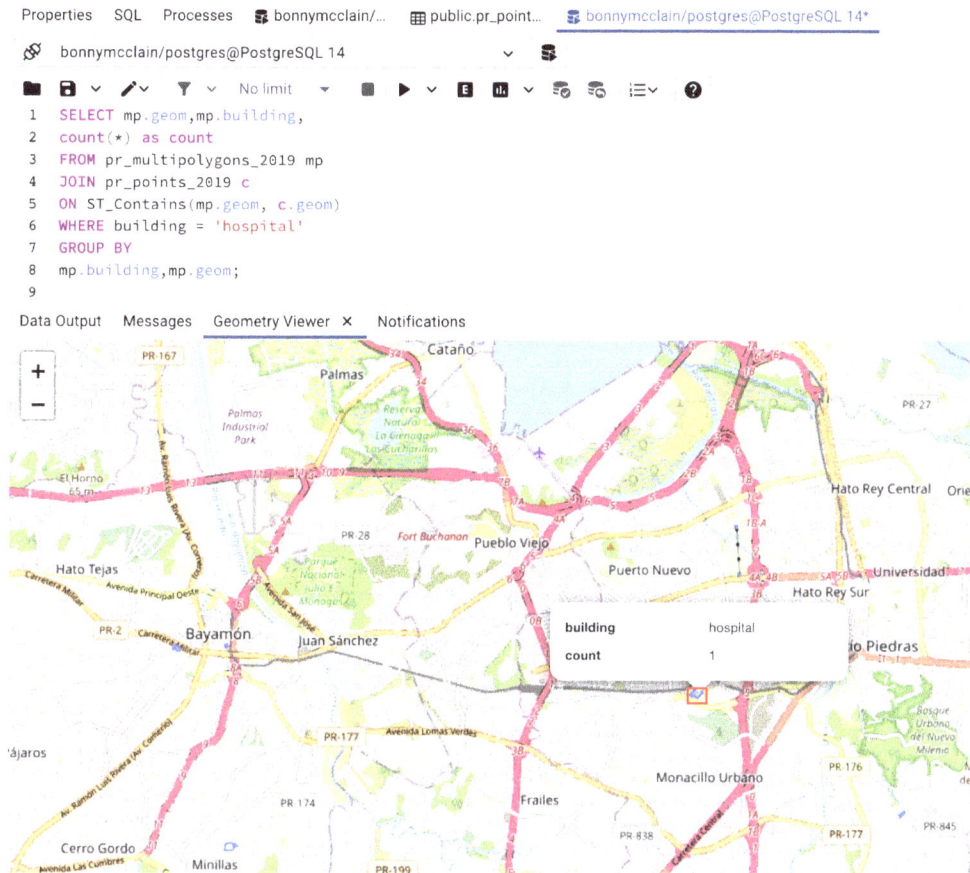

Figure 7.24 – Geometry Viewer is zoomed into an observable polygon representing a hospital in 2019

How can we capture more of the medical facilities that may have been brought online following an earthquake, power outage, or weather disaster such as a hurricane? These are some examples of the type of deeper dives we will take in the next chapter, where we will apply spatial statistics and explore raster data.

## Summary

In this chapter, we dove a little deeper into a few of the tools. You learned how to load a large dataset into QGIS for access within the pgAdmin graphical interface. For me, I think it's a good way to enhance the utility but also the limitations of using pgAdmin as a management tool. I typically use it to explore recently uploaded data via QGIS once I refresh and make sure the data tables are accessible in pgAdmin. Functions can also be run in the query editor and saved to a file to be uploaded into QGIS or your GIS of preference.

We will return to QGIS in the next chapter. Thanks for coming along.

# 8

# Integrating SQL with QGIS

Earlier chapters focusing on SQL and geospatial analysis have expanded your geospatial vocabulary and introduced SQL. You now have access to a host of new tools to explore location intelligence. This final chapter will serve as a segue into an open source workflow for additional learning and insights. The ability to access large datasets and explore them with the expanded functionality of plugins within the QGIS platform, combined with PostGIS and the dynamic spatial query language, is addressed with a higher level of acuity. I suggest downloading the **Long-Term Release** (**LTR**) for the most stable QGIS version: `https://www.qgis.org/en/site/forusers/download.html`.

In this chapter, you will learn to explore SQL queries within DB Manager in QGIS and you will also learn how to import and configure raster datasets along with calculating raster functions.

In this chapter, we'll cover the following topics:

- Understanding the SQL query builder in QGIS
- QGIS plugins
- Working with raster data
- Data resources for raster data

## Technical Requirements

I invite you to find your own data if you are comfortable or access the data recommended in this book's GitHub repository at: `https://github.com/PacktPublishing/Geospatial-Analysis-with-SQL`.

Please find the following datasets for working through examples:

- Geofabrik: `http://download.geofabrik.de/osm-data-in-gis-formats-free.pdf`
- Earth Explorer: `https://earthexplorer.usgs.gov/`

- **Landscape Change Monitoring System (LCMS)** Data Explorer: `https://apps.fs.usda.gov/lcms-viewer/`

- The data for the exercises in all these chapters can be found on GitHub at `https://github.com/PacktPublishing/Geospatial-Analysis-with-SQL`

# Understanding the SQL query builder in QGIS

We have been building SQL queries in DB Manager thus far, but there is one more tool that is useful when writing SQL syntax in large datasets. In the SQL window (visible when you select the icon in the top-left menu), there is a SQL symbol located in the upper-left corner of the console. It opens with a vertical list of SQL prompts.

1. **Columns** lists the actual columns in the database **tables**, which you can select using the dropdown on the far right in the **Data** section. Once you select a table, all of the columns in the dataset are displayed as follows:

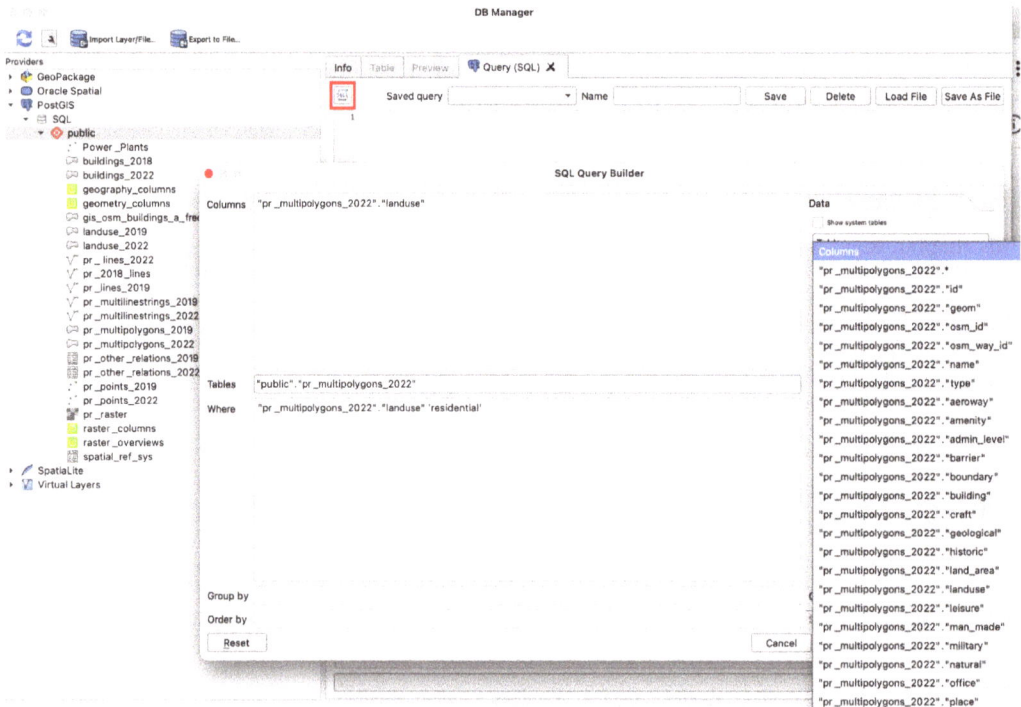

Figure 8.1 – The SQL query builder inside the SQL window in QGI

2.  **Where** is populated when you select **Columns' values** in the right-hand vertical section in *Figure 8.2*. The column value in the data column will populate and display the variables listed in the column of interest. In the example, values for **landuse** equal to **residential** will be available. Also, don't forget to add the = operator.

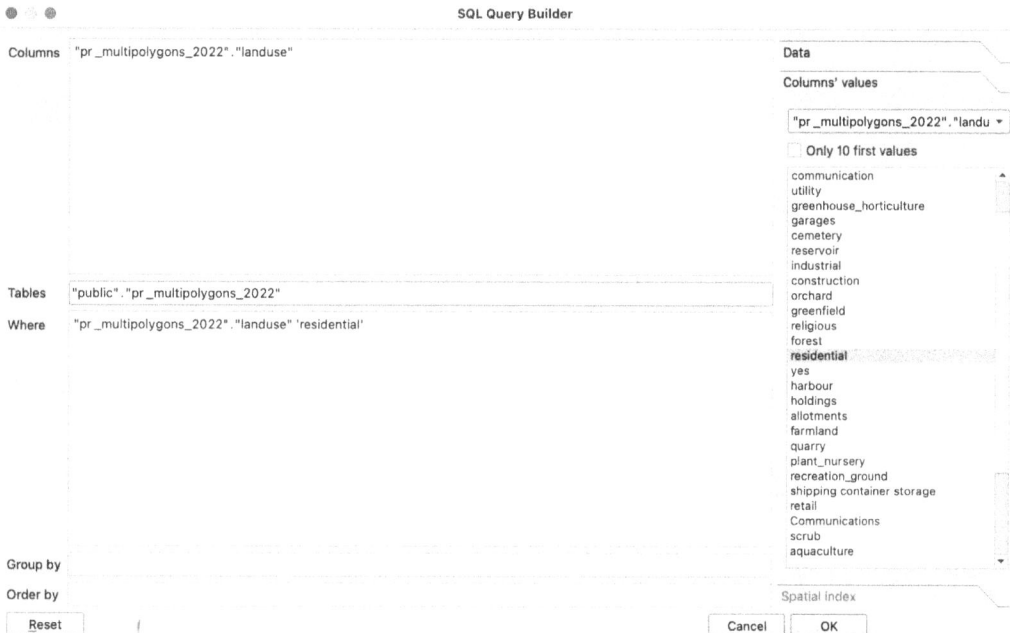

Figure 8.2 – The SQL query builder in QGIS: adding data

3.  The SQL query builder is also a useful tool for improving your skills. When you hit OK, the query populates as shown in Figure 8.3. From here, you can complete the process of selecting the unique identifier in the dataset and the field that contains your **geometry** (**geom**). I suggest naming the layer, especially if you will be querying the dataset numerous times. Otherwise, your layers will be labeled `QueryLayer`, `QueryLayer_2`, and so on. Include `geom` and `id` as unique identifiers for loading the data onto the canvas.

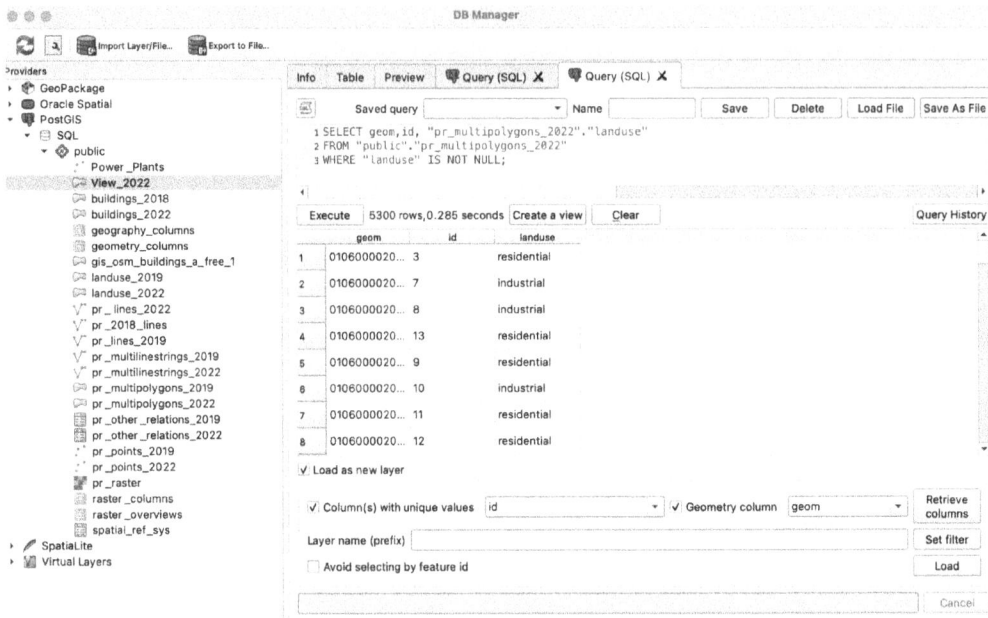

Figure 8.3 – Executing a SQL query in DB Manager QGI

After loading the new layer, right-click on the new layer in the QGIS panel and select **Update SQL**. *Figure 8.4* displays the pane and the option to update the layer on the canvas. Once you zoom in to the layer, the canvas updates with the visual output of the SQL query you entered.

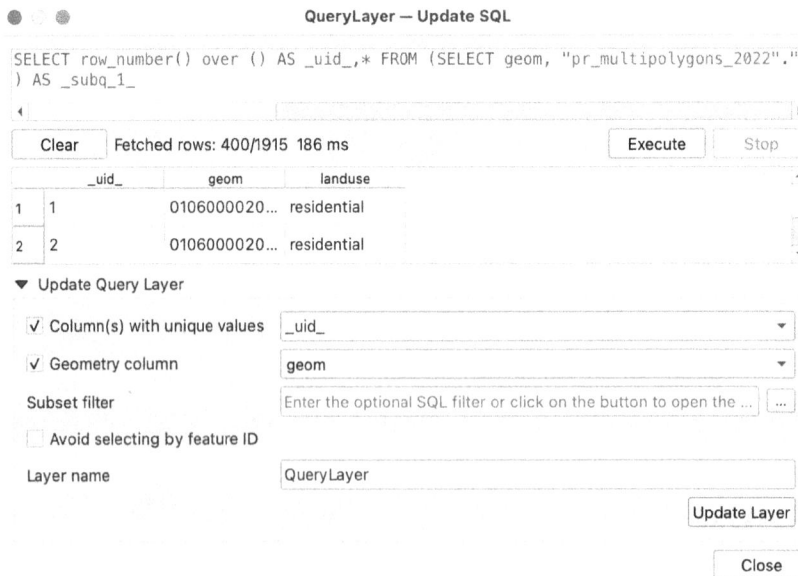

Figure 8.4 – Updating the SQL layer in QGI

The ability to create layers and add or remove them from the canvas yields dynamic interactive queries generated live as you explore datasets from multiple layers. *Figure 8.5* shows `QueryLayer` updated in the canvas.

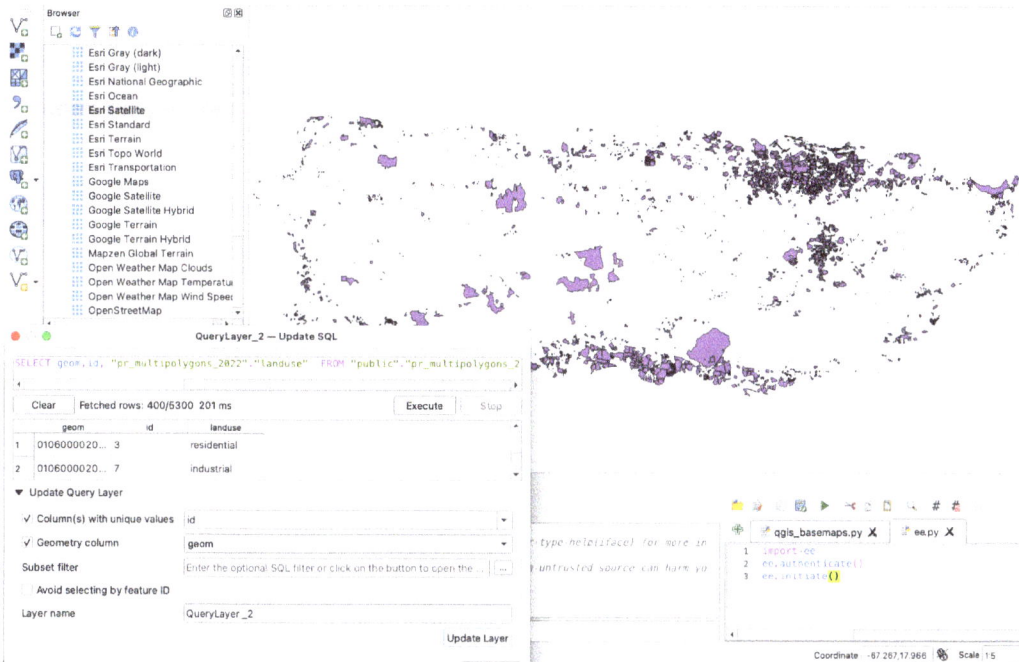

Figure 8.5 – Updated SQL layer on the QGIS canvas

This may be a single step in a several-step workflow or an attempt to visualize the residential properties in the recent 2022 multipolygon dataset. The ability to visualize queries using a robust graphical interface helps to generate deeper insights. For example, how does the number of residential properties compare to the number after the 2017 hurricane, isolated earthquakes, or any other events that have preceded your query?

Changes to our data layers are often visually distinct but let's see whether a SQL query can help us make a quantitative comparison between two different datasets.

## Comparing polygons between two large datasets

Returning to our land use datasets, from Puerto Rico, let's compare the 2022 geometry with that of 2019. Because the data in geofabrik is updated daily, available datasets may differ. The objective of the exercise is to compare data from different time periods. To find archival data locate the link for older files and download to your local computer. Many of these datasets were too large for upload to github. We can build a query that will process each geometry of `landuse_2019`, and then, for each of them (via a lateral join), unite all the intersecting geometries from `landuse_2022` and either retain the entire original geometries or compute the difference. The expectation is that land use can serve as a measure of recovery from the devastating events of Hurricane Maria in 2017 and Hurricane Fiona in 2022 but unfortunately, there have been additional weather events as recently as November 2022 with tropical storm Earl.

Here are the functions we will access to compare polygons between our datasets:

- The ST_Multi geometry (geometry **geom**): https://postgis.net/docs/ST_Multi.html

- The ST_Difference geometry (geometry geomA and geometry geomB): https://postgis.net/docs/ST_Difference.html

- The ST_Union geometry (geometry geomA and geometry geomB): https://postgis.net/docs/ST_Union.html

- The ST_Intersects Boolean (geometry geomA and geometry geomB): https://postgis.net/docs/ST_Intersects .html

Let's examine the query

First, we are using the ST_Multi function to work with this collection of geometries.

The COALESCE function is operating like a filter to return non-null arguments from left to right, as it looks for ST_Difference using the geometry of the landuse_2019 and landuse_2022 datasets.

Run the following statement in the SQL query window in QGIS:

```
SELECT COALESCE(ST_Difference(landuse_2019.geom,
landuse_2022.geom),
landuse_2019.geom) As landuse_2019 FROM landuse_2019 LEFT
JOIN landuse_2022 ON ST_Intersects (landuse_2022.geom,
landuse_2019.geom)
```

You can see the query in the DB Manager SQL window in *Figure 8.6*. Execute and select **Load as new layer**. Select landuse_2019 as the geometry column and load the layer to the canvas.

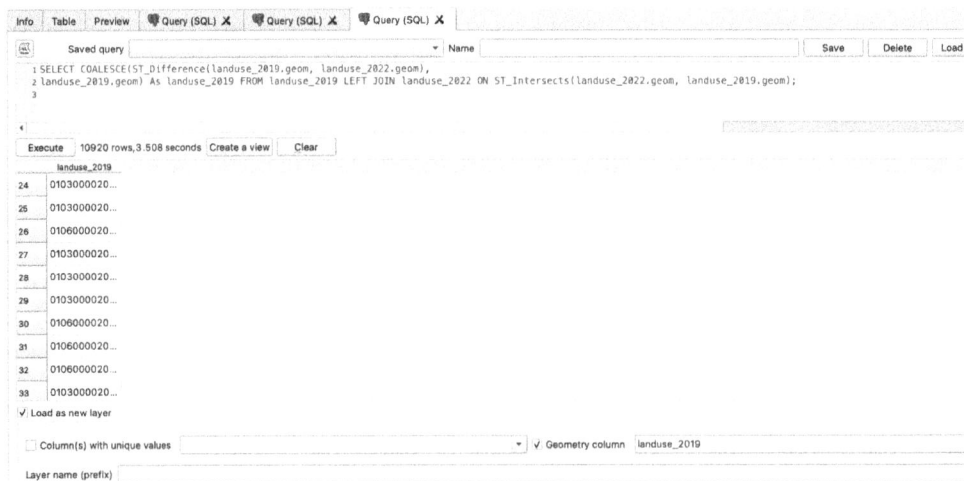

Figure 8.6 – Comparing polygons between two multipolygon data

The polygons are layered in some instances and can be easily masked. To explore SQL queries within the scope of this chapter (and this book), we will look at the differences between the polygons for now and follow it up with a visual comparison in *Figure 8.7*.

Figure 8.7 – Updated query layer in the QGIS canva

Zooming into the layers for `landuse_2019` and `land-use_2022`, you can see the layer styling depicting residential areas as pink, green spaces such as forests, farms, and so on as green, and industrial or military zones as black in *Figure 8.8*.

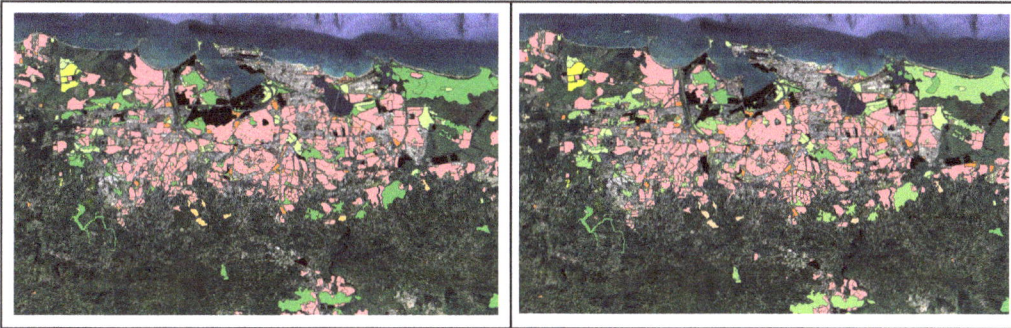

Figure 8.8 – Multipolygon land use from 2019 (left) and 2022 (right)

The biggest areas of change appear to be the return of commercial and residential neighborhoods. Once you detect a change, it is important to focus on a specific area and dig a little deeper. Explore the built infrastructure, and add road networks or other layers to the canvas. See what you notice.

In the following query, which generated *Figure 8.9*, we are exploring both land use datasets and using linear boundaries as a "blade" to split the data. You can also see CROSS JOIN LATERAL here, which is basically a subquery that can go back to earlier entries in a FROM clause and join each row with all of the table functions that apply to that row, the blade geometry:

```
SELECT ST_Multi(COALESCE(
ST_Difference(a.geom, blade.geom),
A.geom
)) AS geom
FROM public.landuse_2019 AS a
CROSS JOIN LATERAL (
SELECT ST_Union(b.geom) AS geom
FROM public.landuse_2022 as b
WHERE ST_Intersects (a.geom, b.geom) ) AS blade
;
```

The *multipolygon landuse_2019* dataset is using linear boundaries as a blade to split the data. ST_Intersects compares two geometries and returns true if they intersect by having, at minimum, one point in common. ST_Union combines the geometry without overlaps, often as a **geometry collection**.

Figure 8.9 – Comparing polygons as a cross join across two datasets using blade geometry

The red areas indicate differences between the datasets. The green polygons are primarily forests and the black color in the map represents industrial or military zones. The polygon in the upper-right corner indicates a change in forest cover in 2022. Notice the polygon is not present in *Figure 8.10* on the right.

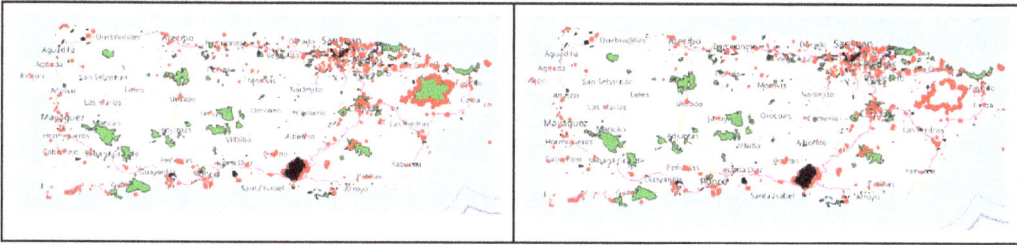

Figure 8.10 – Output of polygon differences between landuse_2017 (left) and landuse_2022 (right)

So far, we have been working with vector geometries. Raster data is a geographic data type stored as a grid of regularly sized pixels. You may have noticed satellite basemaps in GIS interfaces such as QGIS, for example. Each of the cells or pixels in a grid contains values that represent information. Rasters that represent thematic data can be derived from analyzing other data. A common application of raster data is the classification of a satellite image by looking at land use categories.

In the next section, you will be introduced to raster data and learn about a few functions to explore and interact with some of the features in QGIS.

## Raster data

Learning how to work with raster data could actually fill an entire book. My goal is to present an introduction discussing where to find datasets, upload them to your database, and begin interacting within QGIS using a few built-in tools and the SQL query builder.

Navigating to the **Plugins** menu, you should have the DB Manager plugin installed. Let's add the Google Earth and the Google Earth Engine Data Catalog plug-in. You will not need this to follow along with the examples provided in the chapter but it is a powerful tool for locating raster data and my goal is for you to continue exploring SQL and geospatial analysis beyond the pages of this book.

An additional plugin available after you register on the Earthdata site, `https://urs.earthdata.nasa.gov/profile`, shown in *Figure 8.11*, is `SRTM_downloader`.

`SRTM_downloader` is an additional resource for locating raster data. Return to the **Plugins** menu and search for `SRTM_downloader`. Once it's downloaded, enter your credentials on the Earthdata website.

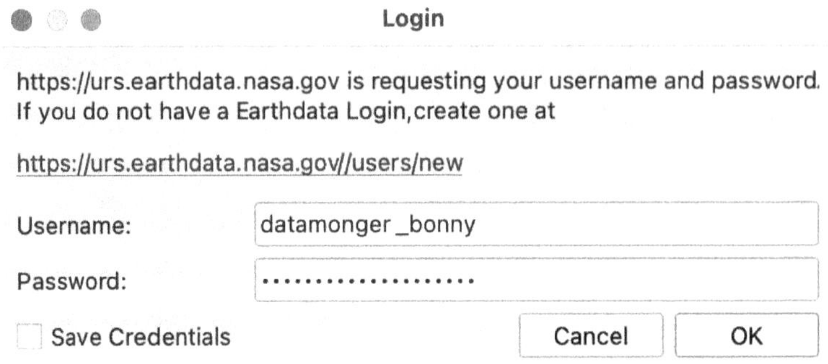

Figure 8.11 – Loading SRTM_downloader in QGIS

On the canvas, select **Set canvas extent**. Click on the icon that was loaded in your toolbar when you downloaded the plugin. SRTM_downloader will appear as shown in *Figure 8.12*. Set the canvas extent to render the raster layer and save the file locally.

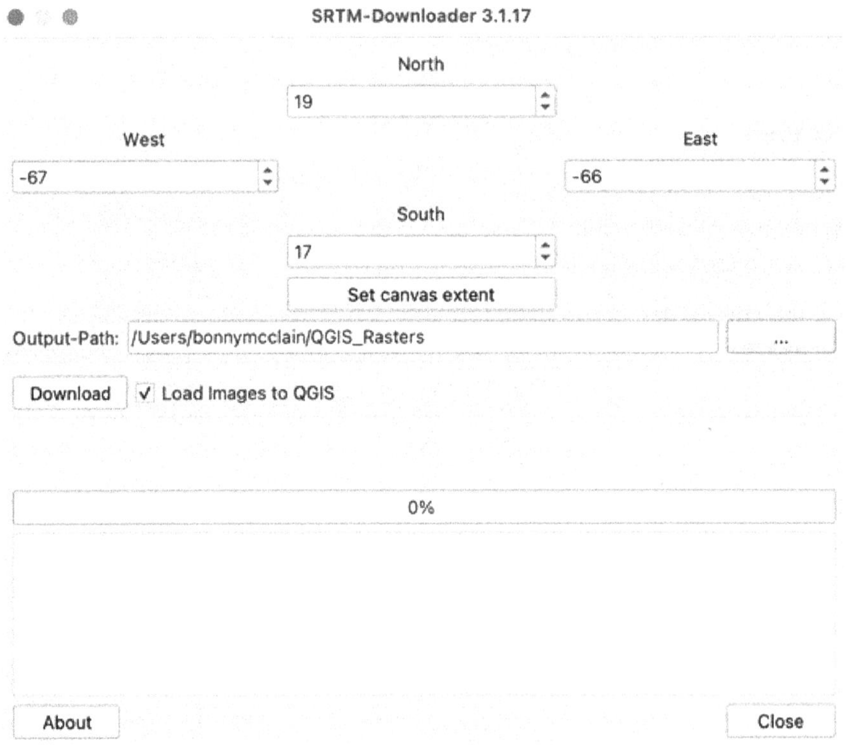

Figure 8.12 – The SRTM-Downloader window

The images will load into the window and directly into your **Layer** panel. The raster uploads into the canvas onto the extent you uploaded into the console in F*igure 8.13*.

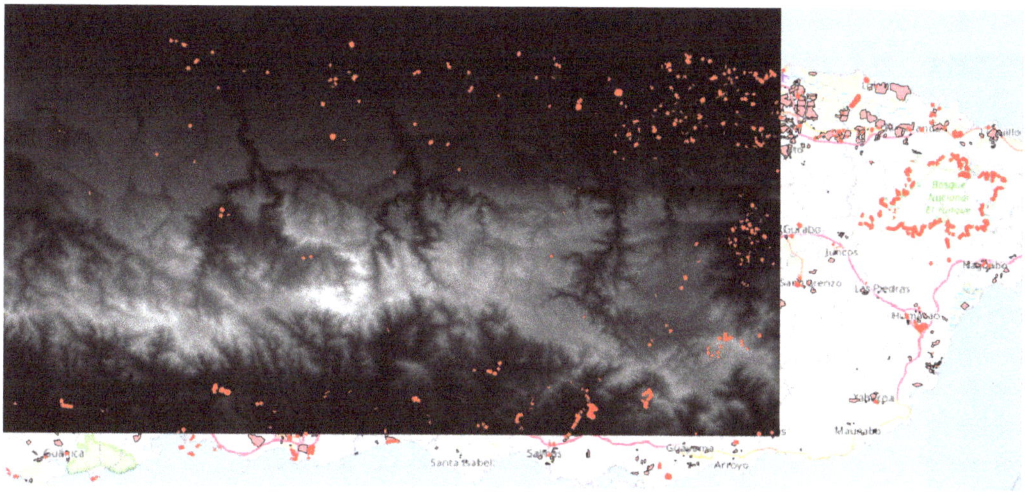

Figure 8.13 – Loading the raster layer from SRTM-downloader

The next set of plugins require slightly more detailed installation instructions.

## Installing plugins

We installed plugins in earlier chapters but a brief reminder of the steps is necessary for a successful installation. The **Google Earth Engine (GEE)** plugin allows you to generate maps directly from terminal or Python console within QGIS. Because access to GEE will need to be authenticated, you will need to link your GEE access to QGIS. You will need an active GEE account (`https://earthengine.google.com/`) and the gcloud CLI (`https://cloud.google.com/sdk/docs/install`) to authenticate within terminal or Python console in QGIS.

*Figure 8.11* displays the **Plugins** option in the top menu bar in QGIS.

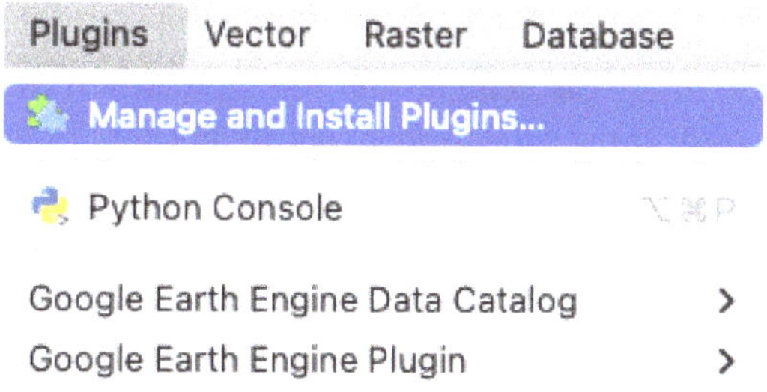

Figure 8.14 – Installing the QGIS plugins

The biggest challenges and sources of error with installing plugins or necessary dependencies result from inaccurate paths. Later in the chapter, we will be importing our data using terminal. An important resource is the **System** setting in QGIS, shown in *Figure 8.15*.

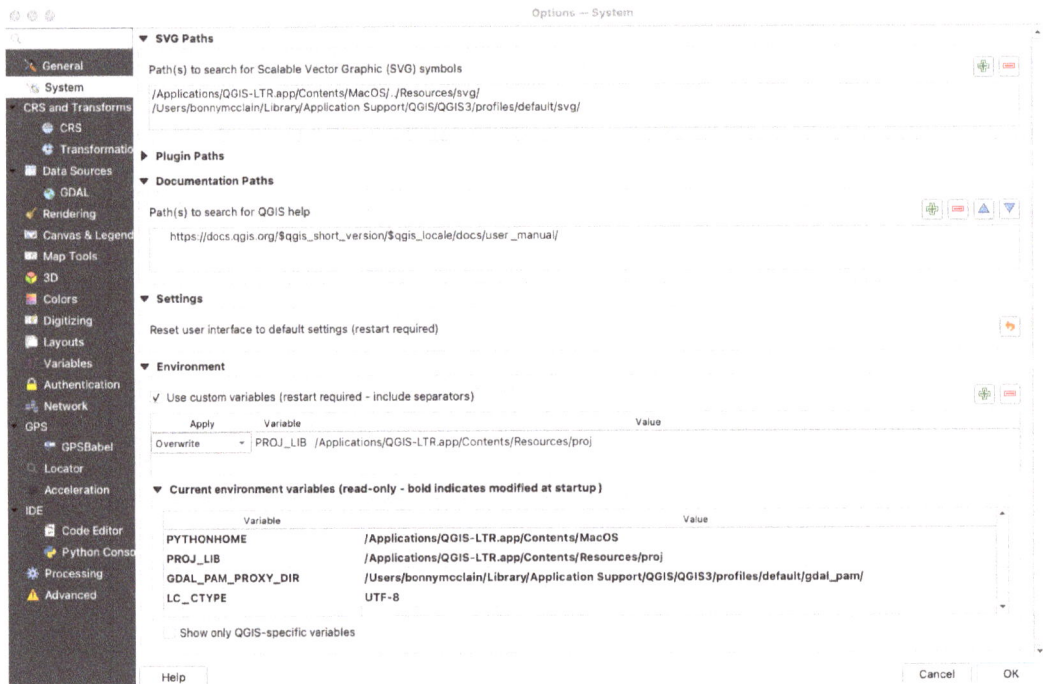

Figure 8.15 – Exploring system paths in QGIS from the menu bar

Let's go through what's happening in the previous figure in the following set of steps:

1.  Click on the QGIS version label in your menu and select Preferences.

2.  If you generate an error, it is important to note the path and the nature of the error. Often, the installation is not in the right location and you need to modify this either manually or by redirecting the path in terminal.

3.  To find terminal window, you can search for it on your computer if it isn't displayed.

4.  Navigate to terminal to install earthengine-api. You can discover the path to your working directory by entering `pwd`, or the path  to your current directory by entering `cd`.

Next, let's cover the QGIS GEE plugin.

## The QGIS GEE plugin

We are installing plugins by using terminal. **Dunder Data** provides a comprehensive guide to installing **Miniconda**, a curated smaller version of **Anaconda**, on your local computer: `https://www.dunderdata.com/blog/anaconda-is-bloated-set-up-a-lean-robust-data-science-environment-with-miniconda-and-conda-forge` Follow these steps:.

1.  Open terminal and write the following command:

    ```
    conda install -c conda-forge earthengine-api
    ```

    Alternatively, if you have `earthengine` installed, run the following update:

    ```
    conda update -c conda-forge earthengine-ap2.
    ```

2.  Returning to QGIS, open your Python console, visible as the Python logo in your menu or plugin (or located by searching in the lower left-hand window) as shown in *Figure 8.13*, and enter the following code into the editor. You will need to access the documentation for installing on other systems or if you used OSGeo4W installer for QGIS.

    ```python
    import ee
    ee.Authenticate()
    ee.Initialize()
    print(ee.Image("NASA/NASADEM_HGT/001").get("title").getInfo())
    ```

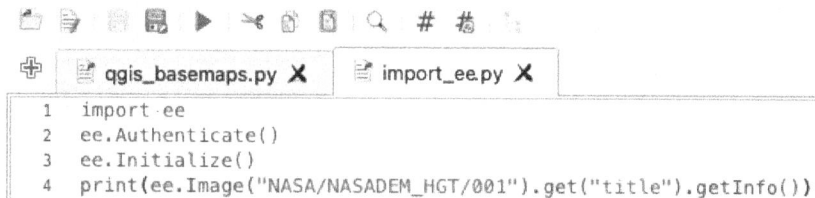

Figure 8.16 – Python console in QGIS

This is  test code to verify that your installation is complete.

There are a variety of system settings and if you run into any difficulties, here are a couple of good resources. Again, the biggest reason for incompatibility is simply that your packages are not in the same path as QGIS:

*   Python installation: `https://developers.google.com/earth-engine/guides/python_install`

*   QGIS Google Earth Engine plugin: `https://gee-community.github.io/qgis-earthengine-plugin/`

One of the biggest advantages of QGIS and GEE integration is direct access to the catalog of scenes and maps. The GEE data catalog also has a plugin to provide direct access to images and collections of satellite data.

## The GEE data catalog plugin

The Earth Engine data catalog is a powerful resource for identifying datasets for exploring raster data. You will need to download the gcloud CLI (`https://cloud.google.com/sdk/docs/install`) to confirm your authentication and access to GEE (`https://developers.google.com/earth-engine/datasets/catalog/landsat`).

I have noticed that with the `QGIS- LTR`, you may need to reinstall the plug-in when logging in again.

Clicking on Database **Toolbar** and selecting the icon for the plugin (seen during plugin download) under **Toolbars** as shown in *Figure 8.17* allows you to visualize the available datasets for your region of interest.

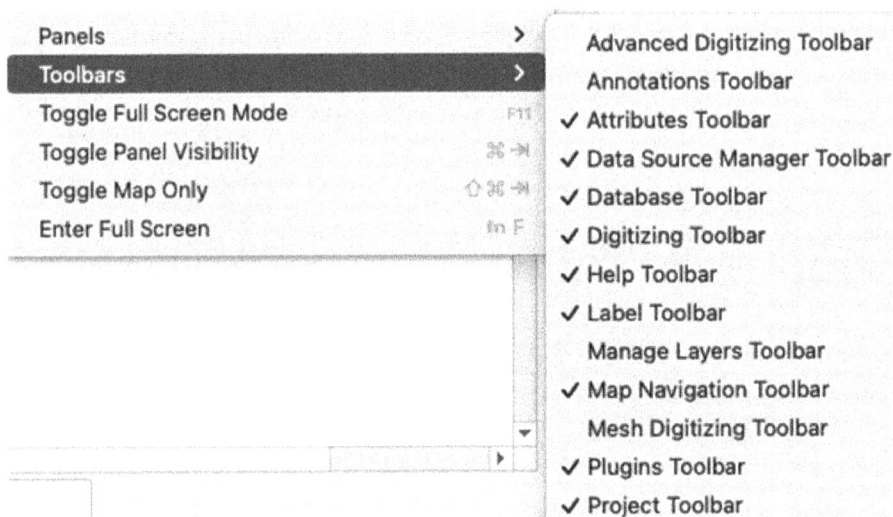

Figure 8.17 – Viewing the toolbars in QGIS

*Figure 8.18* demonstrates that depending on the region you select on your canvas, you can indicate the dataset you want to explore, the bands, the dates for which to view satellite data, and the amount of cloud coverage applicable to your data. This resource is useful for exploring specific date ranges as you select satellite images based on dates. These settings require interaction to find appropriate images. For example, perhaps there were clouds during your date intervals of interest. Allowing more cloud coverage or extending the date of interest will allow more layers for exploration. In addition, each image may not necessarily cover the geographic area you prefer.

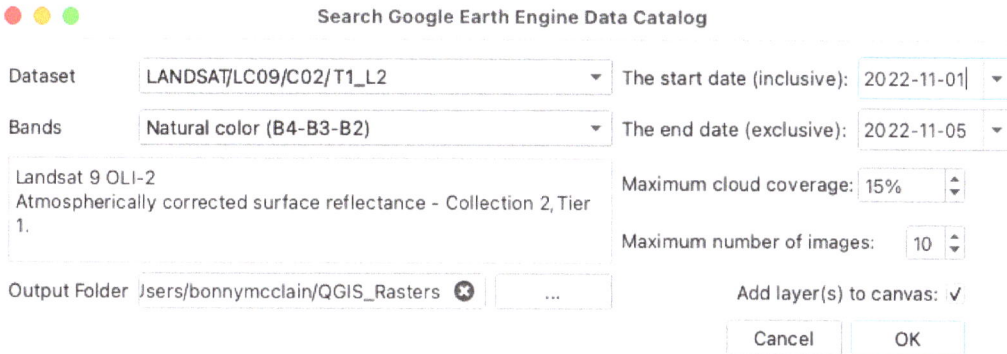

Figure 8.18 – Searching the GEE data catalog within QGIS

The layer is added to the canvas (be sure to check the checkbox in *Figure 8.18*) and although there is cloud cover, I can see areas in *Figure 8.19* into which I can zoom and retrieve useful information. Naturally, if we were only interested in a specific area, we would need to scrutinize the data to locate usable data. Here, we are focused on the land use data from Puerto Rico that we downloaded from Geofabrik in *Chapter 7, Exploring PostGIS for Geographic Analysis*:

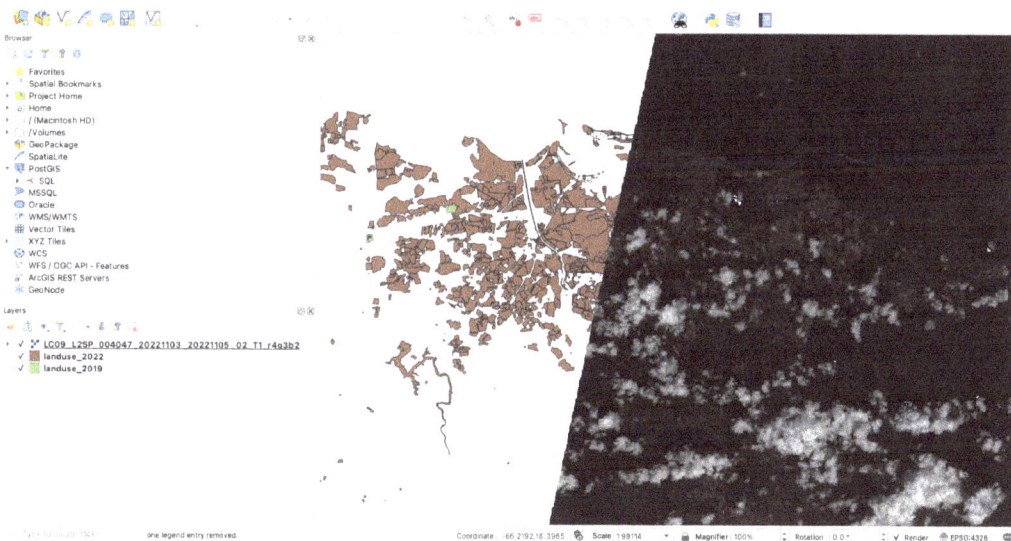

Figure 8.19 – Exploring raster images over an area of interest

Your work with plugins is limited only by your skills, and they are an important tool and resource available within QGIS. There is also a cloud masking plug-in that works with Landsat data for you to explore. I hope this introduction piqued your interest and you continue to explore. Next, I will share a few more out-of-the-box options before introducing you to the actual resource I am using for the examples within this chapter.

## *Other data resources for raster data*

The **Cornell University Geospatial Information Repository** (**CUGIR**) (`https://cugir.library.cornell.edu/`) is a great resource for exploring raster data. The site hosts a wide variety of categories to download data for maps with easy downloads in different formats, as seen in *Figure 8.20*.

Figure 8.20 – CUGIR

For me, CUGIR is a perfect resource for instruction but when I want timely data for a specific question or data exploration, I head over to the **United States Geological Society** (**USGS**) (`https://earthexplorer.usgs.gov/`). Create your free account and you will have access to raster data.

### USGS EarthExplorer

Use the map to select the region of interest or scroll down on the left side to find a feature of interest in the US or the world. In this case, I am interested in Puerto Rico – not a particular area but seeking a sample with minimal cloud cover and compelling topography for our exercise in raster exploration.

Scrolling down through the options, you can select a date range, the level of cloud cover, and your datasets to consider. *Figure 8.21* displays the datasets currently available. I am looking for **Landsat** images.

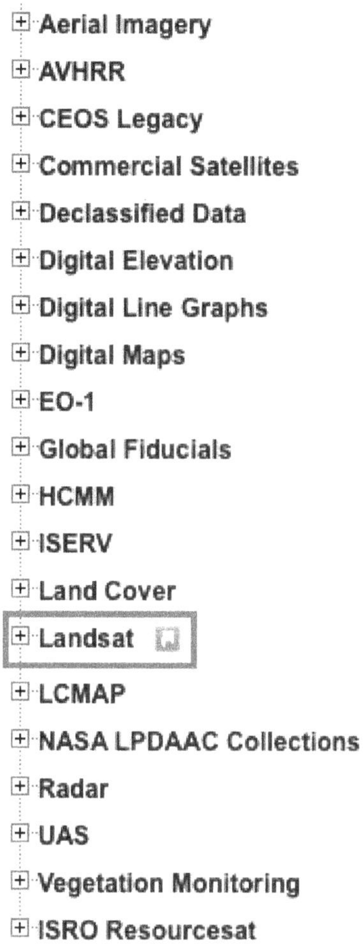

⊞ Aerial Imagery

⊞ AVHRR

⊞ CEOS Legacy

⊞ Commercial Satellites

⊞ Declassified Data

⊞ Digital Elevation

⊞ Digital Line Graphs

⊞ Digital Maps

⊞ EO-1

⊞ Global Fiducials

⊞ HCMM

⊞ ISERV

⊞ Land Cover

⊞ Landsat ▢

⊞ LCMAP

⊞ NASA LPDAAC Collections

⊞ Radar

⊞ UAS

⊞ Vegetation Monitoring

⊞ ISRO Resourcesat

Figure 8.21 – Available datasets in the USGS EarthExplorer hub

Additional search criteria can help you identify the appropriate download. In *Figure 8.22*, the red **Polygon** option that is selected is visible and I opted to select **Use Map** for the reference since I had zoomed in on the location of interest.

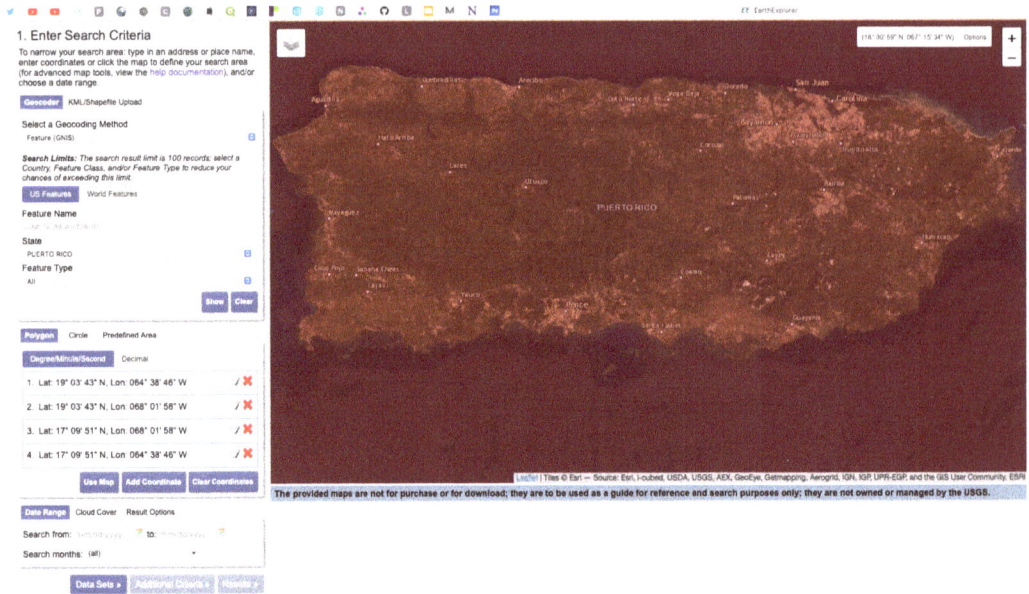

Figure 8.22 – Data search in EarthExplorer

The results are displayed by the date acquired, listed in the left-hand vertical panel. The footprint icon (highlighted in blue in *Figure 8.23*) shows you where the satellite image is located in relation to your highlighted polygon. When you click on the icon next in line, you can actually see what the image will look like. You can see the actual cloud cover of the area of interest, for example. Click on the other images and see how they compare to the other displayed results. Observe the metadata and options for downloading the images to your local computer.

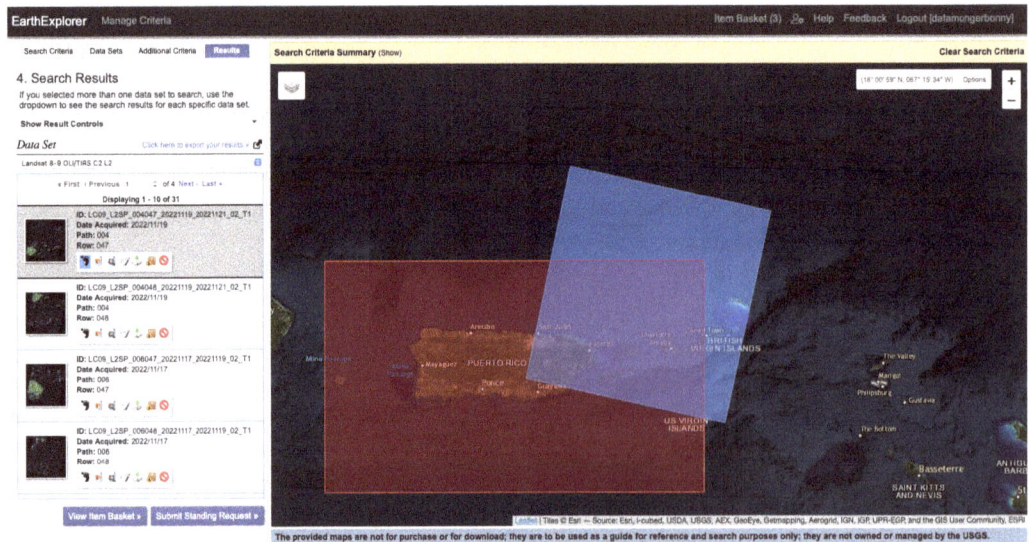

Figure 8.23 – Exploring available images in EarthExplorer

As you browse through images, what you see in *Figure 8.24* displays once you click on the image boundary on the canvas. Scroll down for information about the landsat product you are observing. The **Browse** icon will provide metadata for you to explore. This data is handy if you need to know about how the pixel data is rendered or the projection of the dataset, for example.

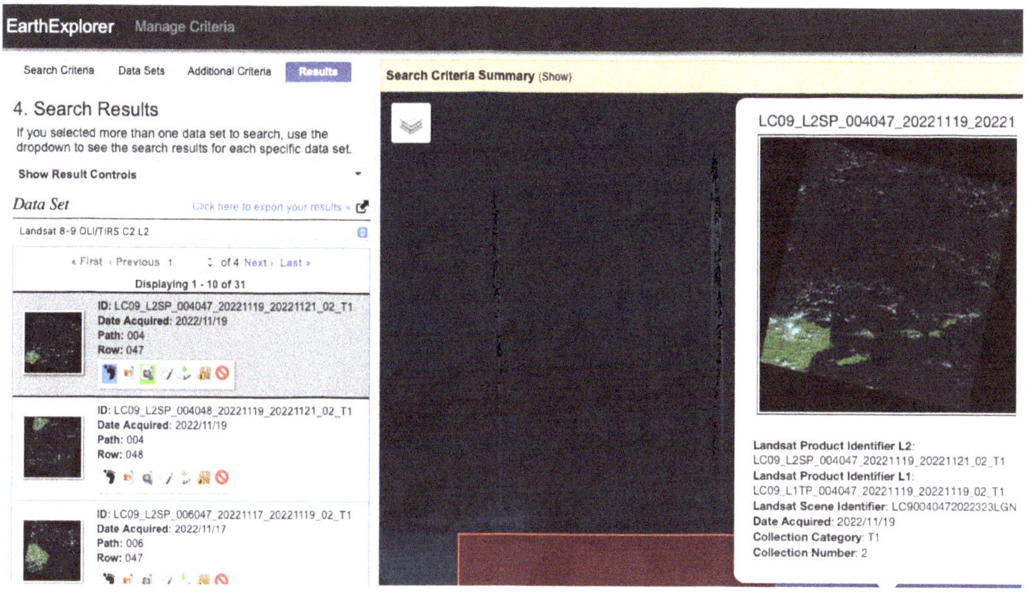

Figure 8.24 – Browsing landsat data in EarthExplorer

I hope you will browse around the examples of different datasets to explore. Many are offered with a variety of download formats, such as **Digital Elevation Models** (**DEMs**), GeoTiff, or even XML (QGIS can convert to GeoTiff).

# Landscape Change Monitoring System

Let's explore the **Landscape Change Monitoring System** (**LCMS**) Data Explorer (`https://apps.fs.usda.gov/lcms-viewer/`), shown in *Figure 8.25*. You can download data from different time periods and observe any changes. The latest data available here was for 2020. Although we can select specific dates from EarthExplorer, the concepts required to process and prepare for introducing spatial concepts in writing SQL queries would be beyond the scope of this book. For example, the workflow I would use professionally involves built-in QGIS raster functions or Python coding.

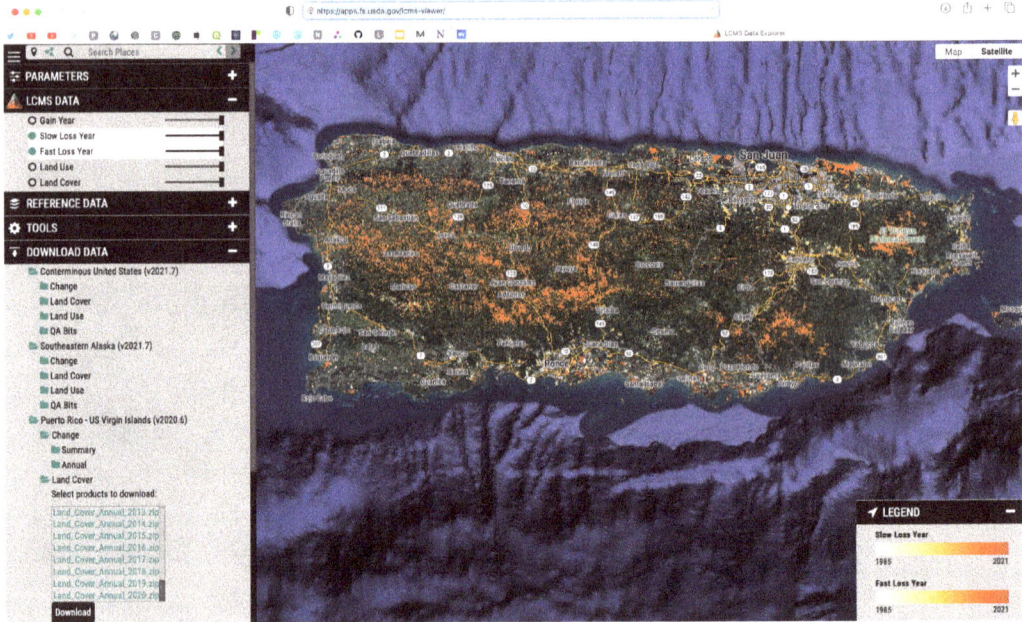

Figure 8.25 – LCMS Data Explorer (Puerto Rico)

For importing raster data and exploring raster data, we can learn about a few SQL functions, but first, we need to locate the data and download it to our local computer. Download the years of interest available by scrolling in the lower-left panel. I selected 2017 and 2020 to explore how land use might have evolved due to a few hurricanes and natural events.

## raster2pgsql

Importing rasters into the database on macOS and even Windows or Linux relies on the raster loader executable, `raster2pgsql`. The **Geospatial Data Abstraction Library** (**GDAL**) supports raster and vector data formats and uploads rasters to a PostGIS SQL raster table.

Open your terminal. You can follow the installation instructions for `postgres.app` here: `https://postgresapp.com/`. When you want to check the installation, simply enter `psql` into the command line.

## Terminal raster import

Copy the following text into terminal and hit *Enter*.

```
psql -c "SELECT PostGIS_Version()"
            postgis_version
```

Your output will show the version number as follows:

```
----------------------------------------
  3.2 USE_GEOS=1 USE_PROJ=1 USE_STATS=1
(1 row)
```

Now that you can be confident that you are up and running, let's import our downloaded raster data. Enter `raster2pgsql -G` to see the GDAL-supported raster formats. Here is a sample.

Run the text in your terminal to see the full list:

```
(minimal_ds) MacBook-Pro-8:~ bonnymcclain$ raster2pgsql -G
Supported GDAL raster formats:
  Virtual Raster
  Derived datasets using VRT pixel functions
  GeoTIFF
  Cloud optimized GeoTIFF generator
  National Imagery Transmission Format
  Raster Product Format TOC format
  ECRG TOC format
  Erdas Imagine Images (.img)
  CEOS SAR Image
  CEOS Image

...
```

The command parameters for importing data are the same, with a few programmable options. The syntax is relatively straightforward once you are familiar.

`raster2pgsql` creates an input file and loads it into 100 x 100 tiles. The `-I` option creates a spatial index on the raster column once the table has been generated. A spatial index will speed up your queries.

The following code statement is explained here. Raster constraints are generated by `-C` and include a `-s` SRID – pixel size, for example. Following the 100 x 100 tiles, you will enter the path to the downloaded file highlighted in the following code. I am using the public schema and have already created a database table (`public.landuse_2017`) and database (`bonnymcclain`) in pgAdmin:

```
raster2pgsql -I -C -s 4139 -t 100x100 /Users/bonnymcclain/
landuse_2017.tif public.landuse_2017 |psql -d bonnymcclain
```

Repeat the sequence for each dataset as follows:

```
raster2pgsql -I -C -s 4139 -t 100x100 /Users/bonnymcclain/
landuse_2020.tif public.landuse_2020 |psql -d bonnymcclain
```

Here is an example from my terminal to demonstrate the output and index creation (a snapshot):

```
(minimal_ds) MacBook-Pro-8:~ bonnymcclain$ raster2pgsql -I -C
-s 4139 -t 100x100 /Users/bonnymcclain/PR_Raster.tif public.
PR_Raster |psql -d bonnymcclain
Output:
Processing 1/1: /Users/bonnymcclain/PR_Raster.tif
BEGIN
CREATE TABLE
INSERT 0 1
INSERT 0 1
INSERT 0 1
...
CREATE INDEX
ANALYZE
NOTICE:  Adding SRID constraint
NOTICE:  Adding scale-X constraint
NOTICE:  Adding scale-Y constraint
NOTICE:  Adding blocksize-X constraint
NOTICE:  Adding blocksize-Y constraint
NOTICE:  Adding alignment constraint
NOTICE:  Adding number of bands constraint
NOTICE:  Adding pixel type constraint
NOTICE:  Adding nodata value constraint
NOTICE:  Adding out-of-database constraint
NOTICE:  Adding maximum extent constraint
 addrasterconstraints
 ---------------------
 t
(1 row)
COMMIT
```

Remember that if you are having trouble, figure out the correct path, then enter cd and the name of the folder to bring you to the right location (depending on where your files are located):

```
cd Applications
```

If you don't indicate a directory, you will simply go to your home folder (or select cd ~). Here are a few more examples:

- ls or list

  You can now see the contents of the folder you selected.

- cd / returns you to the root level of your startup disk.

You might have to look up the shell commands for your specific operating system.

### ST_Metadata

When you want to explore your raster dataset, you can either run a query in pgAdmin or QGIS. I often have both running so I can do simple queries where I am not going to need a graphical interface directly in pgAdmin. The SRID and number of bands are quick pieces of information I like to have access to when writing queries or visualizing maps, as in *Figure 8.26*:

```
SELECT
rid,
  (ST_Metadata(rast)).*
FROM
public.landuse_2017
LIMIT
100
```

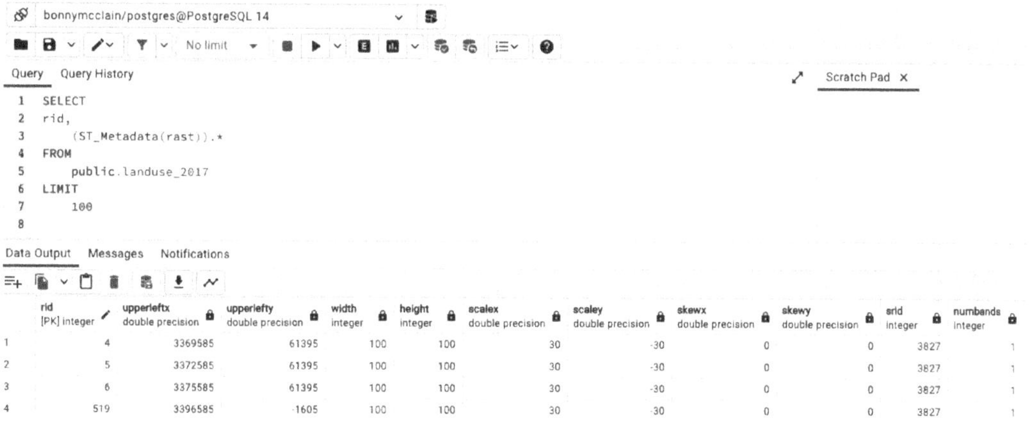

Figure 8.26 – ST_Metadata SQL query

In addition to ST_Metadata running in either pgAdmin or QGIS, the database format in pgAdmin lists raster information in the tree running below the browser, as shown in *Figure 8.27*:

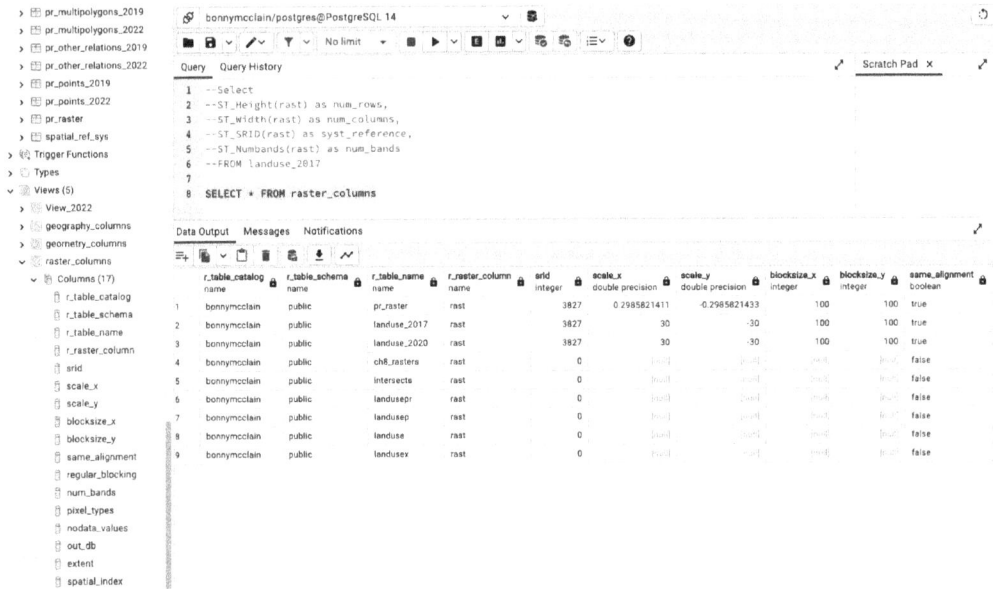

Figure 8.27 – Views in pgAdmin containing the raster_columns information

The SQL code that renders similar information is commented out, so it won't run, indicated by --
If you enter SELECT * FROM raster_columns and scroll from left to right, you will have
information on all of your raster tables.

### Polygonizing a raster

ST_DumpAsPolygons returns geomval rows, created by a specific geometry (geom) and the
pixel value (val). The polygon displayed is from the union of pixels with the same pixel value (val)
from the selected band:

```
setof geomval ST_DumpAsPolygons(raster  rast, integer  band_
num=1);
```

### ST_DumpAsPolygons

Run the following code in the SQL query editor. **Well-Known Text** (**WKT**) refers to a text-based
representation of the geometry:

```
SELECT val, ST_AsText(geom) As geomwkt
FROM (
SELECT (ST_DumpAsPolygons(rast)).*
FROM band1_landuse_2020
WHERE rid = 2
) As foo
```

```
WHERE val BETWEEN 1 and 7
ORDER BY val;
```

Load it as a new layer and now you can view it on the canvas in QGIS, as shown in *Figure 8.28*.

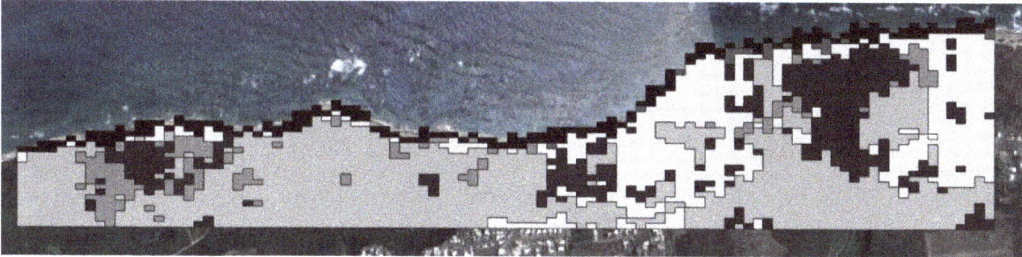

Figure 8.28 – DumpAsPolygons on the northwest border of Puerto Rico

Select the `QueryLayer` or `landuse_2022` polygon to see the rasterized polygon. To increase visibility, select **Singleband pseudocolor** in **Layer Properties**, selecting the defaults in *Figure 8.29*. `QueryLayer` will provide raster information and `landuse_2022` contains information about features such as the military base highlighted in red in *Figure 8.30*.

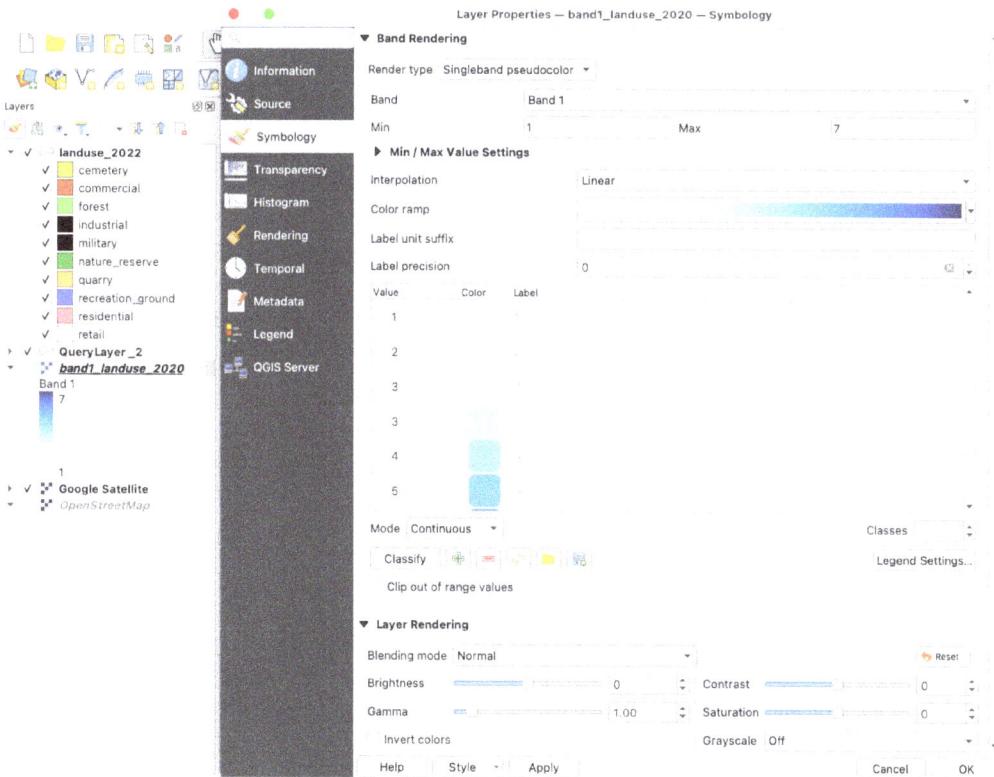

Figure 8.29 – Layer properties in QGIS for raster band rendering

Although I don't recommend going crazy with colors, I have shown these options to illustrate the scope of customization available.

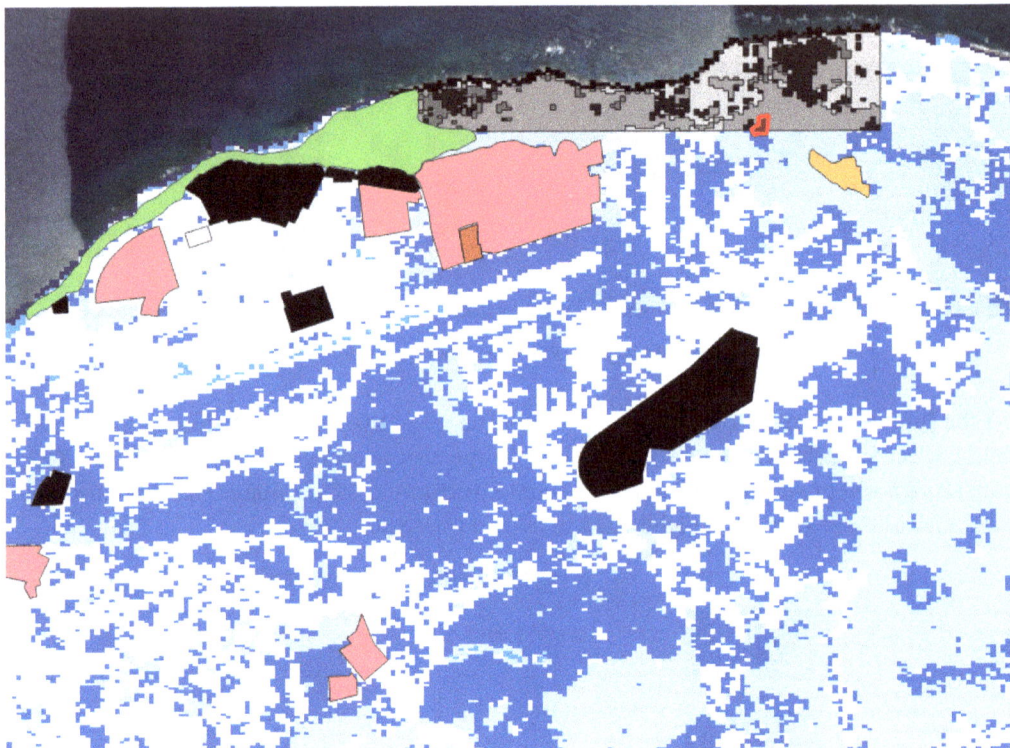

Figure 8.30 – Single-band pseudocolor and land use multipolygons

*Figure 8.31* is a close-up of the information that pops up when you select a polygon on the canvas:

Figure 8.31 – The landuse_2022 feature layer, along with our ST_DumpPolygon and raster layer

You should now have the polygonized raster on the canvas and be able to explore the relationships displayed.

Now, let's summarize the chapter.

# Summary

In this chapter, you saw the value of spatial queries when comparing different datasets consisting of polygons or multipolygons. You uploaded vector and raster data into the database and were introduced to additional QGIS plugins. The ability to access data resources and import them using terminal helped create raster functions and explore the process of creating polygons out of rasters.

In this book, you have been introduced to SQL and geospatial analysis using PostGIS and QGIS as helpful tools in processing both vector geometries and raster data. I recommend exploring additional datasets and adding different functions to your query language skills.

I thank you for the time and attention along the way and wish you luck on your continued spatial journey.

# Index

# U

# V

# W

# ‹packt›

Packtpub.com

Subscribe to our online digital library for full access to over 7,000 books and videos, as well as industry leading tools to help you plan your personal development and advance your career. For more information, please visit our website.

## Why subscribe?

- Spend less time learning and more time coding with practical eBooks and Videos from over 4,000 industry professionals

- Improve your learning with Skill Plans built especially for you

- Get a free eBook or video every month

- Fully searchable for easy access to vital information

- Copy and paste, print, and bookmark content

Did you know that Packt offers eBook versions of every book published, with PDF and ePub files available? You can upgrade to the eBook version at packtpub.com and as a print book customer, you are entitled to a discount on the eBook copy. Get in touch with us at customercare@packtpub.com for more details.

At www.packtpub.com, you can also read a collection of free technical articles, sign up for a range of free newsletters, and receive exclusive discounts and offers on Packt books and eBooks.

# Other Books You May Enjoy

If you enjoyed this book, you may be interested in these other books by Packt:

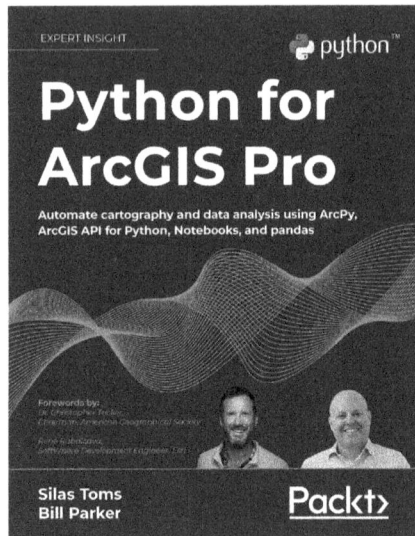

**Python for ArcGIS Pro**

Silas Toms, Bill Parker

ISBN: 9781803241661

- Automate map production to make and edit maps at scale, cutting down on repetitive tasks
- Publish map layer data to ArcGIS Online
- Automate data updates using the ArcPy Data Access module and cursors
- Turn your scripts into script tools for ArcGIS Pro
- Learn how to manage data on ArcGIS Online
- Query, edit, and append to feature layers and create symbology with renderers and colorizers
- Apply pandas and NumPy to raster and vector analysis
- Learn new tricks to manage data for entire cities or large companies

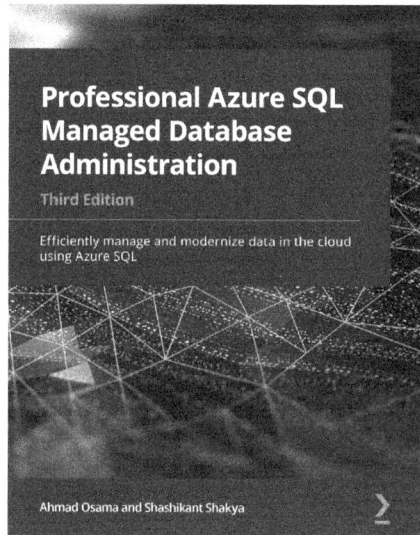

**Professional Azure SQL Managed Database Administration - Third Edition**

Ahmad Osama, Shashikant Shakya

ISBN: 9781801076524

- Understanding Azure SQL database configuration and pricing options
- Provisioning a new SQL database or migrating an existing on-premises SQL Server database to an Azure SQL database
- Backing up and restoring an Azure SQL database
- Securing and scaling an Azure SQL database
- Monitoring and tuning an Azure SQL database
- Implementing high availability and disaster recovery with an Azure SQL database
- Managing, maintaining, and securing managed instances

# Packt is searching for authors like you

If you're interested in becoming an author for Packt, please visit `authors.packtpub.com` and apply today. We have worked with thousands of developers and tech professionals, just like you, to help them share their insight with the global tech community. You can make a general application, apply for a specific hot topic that we are recruiting an author for, or submit your own idea.

# Share Your Thoughts

Now you've finished *Geospatial Analysis with SQL.*, we'd love to hear your thoughts! Scan the QR code below to go straight to the Amazon review page for this book and share your feedback or leave a review on the site that you purchased it from.

`https://packt.link/r/1-835-08314-5`

Your review is important to us and the tech community and will help us make sure we're delivering excellent quality content.

# Download a free PDF copy of this book

Thanks for purchasing this book!

Do you like to read on the go but are unable to carry your print books everywhere?

Is your eBook purchase not compatible with the device of your choice?

Don't worry, now with every Packt book you get a DRM-free PDF version of that book at no cost.

Read anywhere, any place, on any device. Search, copy, and paste code from your favorite technical books directly into your application.

The perks don't stop there, you can get exclusive access to discounts, newsletters, and great free content in your inbox daily

Follow these simple steps to get the benefits:

1.  Scan the QR code or visit the link below

https://packt.link/free-ebook/9781835083147

2.  Submit your proof of purchase
3.  That's it! We'll send your free PDF and other benefits to your email directly

www.ingramcontent.com/pod-product-compliance
Lightning Source LLC
Chambersburg PA
CBHW061411210326
41598CB00035B/6178